EXCEL MANUAL

Peter W. Zehna
Naval Postgraduate School

to accompany

INTRODUCTORY STATISTICS
FIFTH EDITION

&

ELEMENTARY STATISTICS
FOURTH EDITION

NEIL A. WEISS

ADDISON-WESLEY

An imprint of Addison Wesley Longman, Inc.

Reading, Massachusetts • Menlo Park, California • New York • Harlow, England
Don Mills, Ontario • Sydney • Mexico City • Madrid • Amsterdam

To the memory of W. Max Woods

All formulas included on the enclosed diskette are the copyright of Addison Wesley Longman and may not be altered or copied without the written consent of the editor.

Reproduced by Addison-Wesley from camera-ready copy supplied by the author.

Copyright © 1999 Addison Wesley Longman.

All rights reserved. No part of this publication may be reproduced, stored in a retrieval system, or transmitted, in any form or by any means, electronic, mechanical, photocopying, recording, or otherwise, without the prior written permission of the publisher. Printed in the United States of America.

ISBN   0-201-43449-0

1 2 3 4 5 6 7 8 9 10 VG 01009998

PREFACE

The purpose of this manual is to guide the reader in the use of the computer in the solution of statistical problems. It is designed to be used with the computer package Microsoft® Excel 97.† The manual is best used in conjunction with either of the two books, Introductory Statistics, 5/e or *Elementary Statistics 4/e* both by Neil A. Weiss (Reading, Massachusetts: Addison Wesley Longman, Inc., 1999). Indeed, each of the lessons found in this manual are keyed to appropriate sections in these textbooks. Nevertheless, the manual may easily be modified for use with almost any introductory textbook dealing with the standard topics examined here.

One *special* aspect of using the computer for computations is that students can repeatedly process a formula—or several formulas—in a model situation. This approach allows them to see effects of changes immediately. Excel is especially helpful in this regard. First, a set of statistical tools is available in the Data Analysis Tools add-in feature. While these tools cover a wide variety of routines, a few standard ones are missing and there is no control over the output for adjusting to individual tastes.

However, the wealth of functions available for use in individual cells, coupled with extensive formatting capability, is versatile enough to allow you to create nearly all of the standard routines found in an introductory statistics textbook to suit your own purpose. We exploit this feature of Excel extensively and have provided four workbooks with this manual to coordinate the Excel tools with other desirable outputs. Alter these at your own risk for Excel is fragile in the sense that it is extremely easy to inadvertently write over a cell and destroy its contents. We strongly recommend that you keep the original copy in a safe place and work with a copy downloaded to your hard disk.

The author would like to thank Neil Weiss for providing an excellent textbook to write supplementary material for and also for his generosity and support in supplying the necessary TEX macros as well as appropriate files and the use of various examples found in the textbook itself. A debt of gratitude is extended to all the staff at Addison Wesley Longman with special mention of Laura Potter for her patience, understanding, and rapid response to concerns.

Monterey, California   P.W.Z.

---

† Hereinafter referred to as simply Excel.

TO READERS OF ELEMENTARY STATISTICS/4e ONLY

All of the material in *Elementary Statistics/4e* (*ES/4e*) is included in *Introductory Statistics/5e* (*IS/5e*); however, the converse is not true. This manual has been written with *IS/5e* as the primary reference point for coordinating chapters of the textbook with lessons in the manual. Hence, for readers of *ES/4e*, the following table gives cross references from the logo appearance in your textbook to the corresponding page number reference in this manual.

Logo Appearance in Textbook *ES/4e*	Corresponding Page Number Appearance in this Manual
Chapter 4	Lesson 14
Section 5.1	Section 4.1
Section 5.2	Section 4.2
Section 5.3	Section 4.3
Section 5.4	Section 5.1
Section 5.5	Section 5.2
Section 5.6	Section 5.3
Section 9.4	Section 9.5
Section 9.5	Section 9.6
Section 10.4	Section 10.5
Chapter 11	Lesson 12
Chapter 12	Lesson 13
Chapter 13	Lesson 16
Chapter 14	Lesson 15

CONTENTS

LESSON 1
GETTING STARTED   1

1.1 Introduction   2
1.2 Excel workspaces and menus   3
1.3 Input/Output   8
1.4 Excel files   12
1.5 Simple random sampling   14
1.6 Systematic random sampling   16

LESSON 2
ORGANIZING DATA   19

2.1 Grouping data   20
2.2 Graphs and charts   24

LESSON 3
DESCRIPTIVE MEASURES   30

3.1 Measures of center   31
3.2 The sample mean   33
3.3 Measures of variation; the sample standard deviation   35
3.4 The five-number summary   37
3.5 Descriptive measures for populations   41

LESSON 4
PROBABILITY CONCEPTS   43

4.1   Probability basics   44
4.2   Contingency tables; joint probability distributions   45
4.3   Conditional probability   50

LESSON 5
DISCRETE RANDOM VARIABLES   53

5.1   Discrete random variables and probability distributions   54
5.2   The mean and standard deviation of a discrete random variable   58
5.3   The binomial distribution   59
5.4   The Poisson distribution   62

LESSON 6
THE NORMAL DISTRIBUTION   67

6.1   Normally distributed random variables   68
6.2   Areas under the standard normal curve   70
6.3   Working with normally distributed random variables   73
6.4   Normal probability plots   74

LESSON 7
THE SAMPLING DISTRIBUTION OF THE MEAN   79

7.1   Sampling distributions; sampling error   80
7.2   The sampling distribution of the mean   81

Contents vii

LESSON 8
CONFIDENCE INTERVALS FOR ONE POPULATION MEAN 89

8.1 Confidence intervals for one population mean; σ known 90
8.2 Confidence intervals for one population mean; σ unknown 92

LESSON 9
HYPOTHESIS TESTS FOR ONE POPULATION MEAN 96

9.1 Some preliminaries 97
9.2 One-sample z-test for a population mean 99
9.3 One-sample t-test for a population mean 102

LESSON 10
INFERENCES FOR TWO POPULATION MEANS 106

10.1 Inferences for two population means (σs equal) 107
10.2 Inferences for two population means (σs not equal) 109
10.3 Inferences for two population means using paired samples 112

LESSON 11
INFERENCES FOR POPULATION STANDARD DEVIATIONS 116

11.1 The χ^2-distribution 117
11.2 Inferences for one population standard deviation 119
11.3 Inferences for two population standard deviations 123

LESSON 12
INFERENCES FOR POPULATION PROPORTIONS 128

12.1 Confidence intervals for one population proportion 129
12.2 Hypothesis tests for one population proportion 131
12.3 Inferences for two population proportions using independent samples 132

LESSON 13
CHI-SQUARE PROCEDURES 136

13.1 The χ^2-distribution 137
13.2 Chi-square goodness-of-fit test 138
13.3 Contingency tables; association 141
13.4 Chi-square independence test 145

LESSON 14
DESCRIPTIVE METHODS IN REGRESSION AND CORRELATION 149

14.1 The regression equation 150
14.2 The coefficient of determination 153
14.3 Linear correlation 156

LESSON 15
INFERENTIAL METHODS IN REGRESSION AND CORRELATION 157

15.1 The regression model: analysis of residuals 158
15.2 Inferences for the slope of the population regression line 162
15.3 Estimation and prediction 165
15.4 Inferences in correlation 167

LESSON 16
ANALYSIS OF VARIANCE (ANOVA) 170

16.1 The F-distribution 171
16.2 One-way analysis of variance 171

APPENDIX
ANSWERS TO PROBLEMS A-1

INDEX I-1

LESSON 1

Getting Started

GENERAL OBJECTIVE In this lesson we introduce you to some preliminary material that is necessary for utilizing Microsoft® Excel 97. You will learn how to access Excel and how to use this manual for solving problems in the two books *Introductory Statistics, fifth edition* (hereinafter referred to as *IS/5e*) and *Elementary Statistics, fourth edition* (hereinafter referred to as *ES/4e*), both by Neil Weiss.[†] You will learn how to enter data, edit the data for corrections, and execute Excel's extensive function capability.

LESSON OUTLINE
1.1 Introduction
1.2 Excel workspaces and menus
1.3 Input/output
1.4 Excel files
1.5 Simple random sampling
1.6 Systematic random sampling

[†] Published by Addison Wesley Longman, Inc., Reading, MA.

1.1 Introduction

In this supplement you will learn how a computer can be used to solve statistics problems. The most commonly used programs for statistical work are taken from *statistical software packages*. Statistical software packages are collections of statistical computer programs written by some organization or individual. While not specifically a statistical software package, Excel contains a vast number of statistical routines and functions that can be applied to perform the statistical analyses discussed in the books *IS/5e* and *ES/4e* by Neil A. Weiss. To get the most from this supplement, you should use it while at the computer and do the examples and problems as you read.

For ease of reference, we will call this supplement "the manual" and either textbook, *IS/5e* or *ES/4e*, "the book." The manual is divided into *lessons*. We have used the word "lesson" instead of "chapter" to avoid confusion with the book. The first four sections of this lesson, Lesson 1, can be read immediately. They do not require any knowledge of statistics.

Most lessons in the manual correspond to chapters in the book having the same number. That is, Lesson 2 of the manual discusses how Minitab can be applied to solve the statistical problems considered in Chapter 2 of the book; Lesson 3 discusses using Minitab to solve the statistical problems considered in Chapter 3 of the book; and so on. Exceptions occur in distinguishing between the two books *IS/5e* and *ES/4e*. We will now explain how we accommodate that distinction.

Look on page 14 of this manual. To the right of the section heading "1.5 Simple random sampling" you will see a box with numbers E-29 : I-29 inside of it.† This means that you should begin that section of the manual only when you have read to page 29 of the corresponding book. On page 29 of the book you will find in the left margin at the end of the example a symbol (an Excel logo) like the one appearing here in the left margin.

This symbol serves two purposes. First, it signifies that the material just covered in the book is discussed in the manual. Second, it indicates the exact place to which you should read on page 29 of the book before commencing with the material in the manual where the boxed page numbers occur.

We assume the reader is familiar with the basics of using Windows 95 and Excel 97 on the computer. This includes the use of a mouse as well as familiarity with menus, dialog boxes and the general manipulation of windows. Excel reference manuals run as high as 1,000 pages and more. Clearly, this introductory chapter can only highlight those aspects of Excel that we will be exploiting the most in this manual. We can only touch the surface of the extensive capabilities of this program and invite you to use the excellent online help facility found within the program.

---

† E for *Elementary Statistics* and I for *Introductory Statistics*; eventually, the page numbers will be different. Also we would call your attention to the legend given on the inside front cover of this manual which map those sections of *Elementary Statistics* which differ from those in *Introductory Statistics*. When the references refer to different pages in the two books covered by this manual, you will see the difference in the appropriate box.

1.2 Excel workspaces and menus

Once you have accessed Excel, the computer will display an opening page something like that in Figure 1.1. This page will vary somewhat according to the implementation you are using.

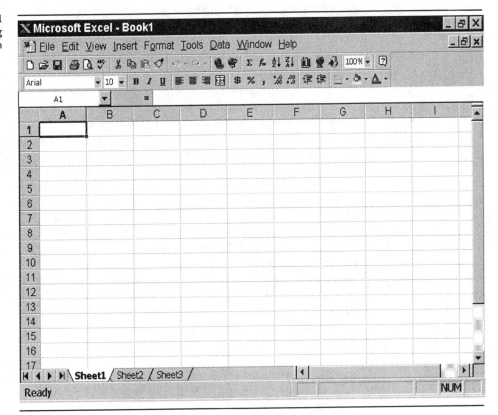

FIGURE 1.1
Opening
Excel Screen

At the top of opening screen is the *title* bar, a feature common to all Windows applications. It identifies the program as Microsoft Excel, with its standard logo in the left corner. In the right-hand corner of the title bar you will find the common Windows control buttons, ▬ ◻ ✕. The first one is used to *minimize* the window (reduce the window to an icon in the Windows taskbar shown at the bottom of the screen). The second, the *maximize* button, causes the window to fill the entire screen for maximum visibility (as shown here). The picture in the icon then changes to a double window ▣ and this modified button *restores* the window to the size it was before being maximized. Finally, the last button, ✕ , in any Windows titlebar is the standard Windows button for *closing* the active program.

Below the title bar is the typical Windows *menu* bar, this one associated with the Excel program. As always, a click with the mouse on any of these (or pressing the [Alt] key along with the underlined letter) will produce a drop down menu from which other choices can be made. We will exploit these menus in many of the applications to follow.

The next two lines consist mainly of small icons with special functions and we collectively call these the *toolbar*. Following the toolbars you find two *text boxes*, spaces for entering material from the keyboard. The second, rather long text box is called the *formula bar* in particular and here is where we can input formulas from the extensive list of those available. The text box on the left is called a *name* box and is useful for naming ranges of cells for subsequent use. We will have more to say about this in the next section on Input/output.

WORKSPACES

That brings us to the general workspace area where we will do all of the work needed to implement the procedures in *IS/5e* and *ES/4e*. This space is called a *workbook* and this empty one is named **Book1** as you can see in its own title bar (complete with its own set of Windows control buttons). At the bottom of the workbook you will find three tabs marked, respectively, **Sheet1**, **Sheet2**, **Sheet3**. These are called *worksheets* in Excel and each workbook comes as a default with three empty sheets, although you may certainly add more. In the open worksheet you see the typical layout of a spreadsheet in cells, labeled A,B,C,... across the top (256 in all) and 1,2,3,... down the left side (65,536 in all). This allows you to give each of the 16,777,216 cells in the worksheet a unique address, like **K4** according to the column (K) and row (4) of its location.

The active cell is indicated with a heavy border – in this case it is cell **A1** (also indicated in the name box as the current name for this cell). You may change the active cell by pointing to the cell with the mouse pointer (an open cross ✥ in Excel) and clicking with the mouse. In addition, you may navigate the sheet by using the direction arrow keys, [↓], [↑], [←], and [→] on your keyboard.

On the right-hand side of the worksheet you find the standard Windows vertical scrollbar allowing you to scroll the sheet vertically with the two scroll buttons, or more rapidly by holding the scroll bar between the buttons with the mouse pointer and dragging it up and down. At the bottom of the worksheet, next to the sheet tabs you find the standard Windows horizontal scrollbar allowing you to scroll the sheet horizontally with the two scroll buttons, or more rapidly by holding the scroll bar between the buttons with the mouse pointer and dragging it left and right. Right next to the left scroll button is the *Tab split bar*; grab this with the mouse and drag left or right to change the amount of display space for Worksheet tabs.

Finally, just below the sheet tabs you will find the *Status bar* which is reserved for various messages and to indicate the status of certain keys. For example, in Figure 1.1 we see that the worksheet is in the **Ready** state and that the numeric keypad on your computer is active as indicated by [NUM]. The last horizontal bar is the standard Windows *Taskbar* and always has the start button, [🏁 Start], first. There is also an indication that **Microsoft Excel** is active since its arrival in the Taskbar means the program must have been minimized. Of course your Taskbar may differ from the one that appears here.

MENUS

The menu bar gives you a list of choices for various Excel features. In Figure 1.1 observe that there are nine menu choices in the Excel window. These are displayed as **F**ile , **E**dit , **V**iew , **I**nsert , F**o**rmat , **T**ools , **D**ata , **W**indow , and **H**elp , with one letter of each underlined. You may activate any one of these menus from the keyboard by pressing the [Alt] key at the same time as the underlined letter. An example of this would appear in the manual as [Alt]+[F]. We call these menu choices *Excel commands* and they are our primary focus in this manual.

With a mouse connected to your system, you may access, or *select*, these menus by moving the mouse pointer to the desired menu option and tapping the left mouse button. This action is referred to as *clicking on* the menu name. In all of the cases where we give menu instructions in this manual, this will be the primary instruction for selecting an item. There is almost always an alternative keyboard key that you may press instead and this will always be understood if not explicitly mentioned.

Either way, selecting a menu brings up a screen with several topics called *submenus*. This is illustrated in Figure 1.2 where we elected the F**o**rmat menu.

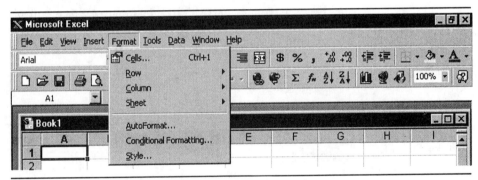

FIGURE 1.2 Format drop down menu with submenus

These submenus can be accessed by clicking on the topic with the mouse, by pressing the underlined key (without the [Alt] key), or use the direction arrow keys to highlight the topic, and then press the [Enter↵] key. See Figure 1.3.

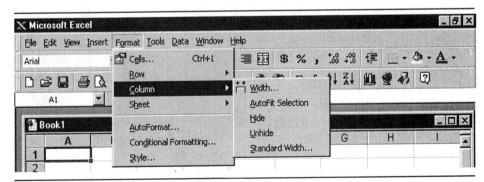

FIGURE 1.3 Format menu with Column submenu

Some of the submenus have a right-pointing triangle, ▶, to the right of its name indicating that other submenu choices are available when it is selected. In Figure 1.3, for example, we show the submenu that appears when you select **Column** from the menu choices for **Format**. Other choices have ellipses dots,..., and this indicates that a *dialog box*, a menu selection that requires further keyboard input from you, will appear when selected. We will be demonstrating many of these matters in the material to follow. Consult the Excel User's Guide for additional details.

We do not always discuss every option for selecting a menu item but use the general direction "select," leaving the options up to you. An example of a menu selection as it will appear in this manual is the following.

Excel commands:

1 Choose **Format** ▶ **Column** ▶ **Standard Width**...

This sequence means: First select the **Format** menu (by whatever means) *and then* select the **Column** submenu (by whatever means) *and then* select the **Standard Width**... dialog box (by whatever means). The symbol ▶ is used as a separator between the choices and can be read *"and then select."* We will number the steps for ease of use and future reference, particularly when many steps are involved.

To return to dialog boxes, Figure 1.4 displays the dialog box that appears when you execute the following simple command.

Excel commands:

1 Choose **Edit** ▶ **Replace**...

FIGURE 1.4
The Replace dialog box in Edit

Dialog boxes contain other boxes, called *text* boxes that require input from you, the user, in order to effect a particular command. In this example, the first text box, labeled **Find what:** is where you will type an expression you wish to find in the open file. Below that is a text box labeled **Replace with:** and awaits your input for an expression to replace the one you just entered. The **Search:** text box has a select button (with a down triangle) ▼ that will bring up a drop down list for your selection of searching by either rows or columns. This relieves you from having to type that selection. Next to this box are two small squares, ☐, called *check* boxes. When you click on or select one of these, a check mark will appear if one is not there, ☑, and disappear (we call that *deselect*) if one is there.

We will illustrate these and other boxes many times as we proceed in the manual. The mouse pointer and the [Tab] key are particularly useful in getting around different parts of a dialog box. From now on, when we say "Press [Tab] to arrive at such and such a place in the dialog box (usually given by name)" we mean press [Tab] a sufficient number of times to arrive at that place or else point the mouse there and click. Pressing [Shift]+[Tab] reverses the order of traversing in a dialog box. The directional arrow keys may also be used along with the mouse pointer to move among various places in the dialog box. In any particular application, we give you explicit instructions on which options to select and what input to enter.

As a final example for this topic, suppose you execute the following command.

Excel commands:

1 Choose <u>W</u>indow ➤ <u>A</u>rrange...

You will see the submenu depicted in Figure 1.5 (ignore the meaning for now).

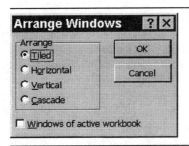

FIGURE 1.5
The Arrange dialog box in Window

Here, the text box has been replaced by a selection of *option* (also called *radio*) buttons, ○, in an area or *field* named **Arrange**. You *select* one of these by clicking on it with the mouse (or pressing the underlined letter without the [Alt] key) in which case it changes to ⊙.[†] There is also a check box, ☐, under these which you select and deselect by clicking on it or pressing the underlined key on the keyboard.

Finally, you see two closing buttons, [OK] and [Cancel]. When you click [OK] (or else press [Tab] to arrive at the [OK] button and press the [Enter] key), Excel generates the result, called an *output*, corresponding to the options you selected. When you click [Cancel] (or else press the [Esc] key) the execution is canceled entirely.

Here are some typographical conventions that we use throughout the manual. In referring to Excel prompts and certain other visuals *as you would see them on your screen*, we will use a typeface like **Group** and **Outline**. Directions to you, the reader, to type a command will be displayed as: Type <u>Normal</u>. You type the material above the underline *and then* press the [Enter] key. We use the typeface **File Name** to refer to various Excel commands and other computer visuals *outside the context* of a display you would see on the screen.

Thus, for example, we use CORRELATION when referring broadly to this Excel command. If we were directing you to a screen output line in which this command

---

[†] You may also use the [↑] and [↓] keys on the keyboard to navigate these choices.

occurred, we would refer to its occurrence in that display as **CORRELATION**. On the other hand, if we are directing you to type this command, you will see the instruction: Type `Correlation`. If we forget to mention it explicitly, it will be understood that you follow this by pressing the [Enter↵] key.

1.3 Input/output

In Excel, Data are typically entered in contiguous cells, called a *range*. In this manual such a range will almost always be a set of contiguous rows in a given column, say from cell A1 through cell A10. This range is denoted **A1:A10** and the data in those cells may always be addressed in that worksheet by that range. However, as Excel points out, it is a lot easier to remember a name than a range. We concur, and all of the data sets in this manual have been given a name according to the context in which the data are used in *IS/5e* and *ES/4e*. Thus, for the purposes of this manual, you will rarely have to enter data yourself. Nevertheless, you should become acquainted with some data entry fundamentals for the entry of data other than the examples treated in this manual.

Basically you enter data, both numeric and character (or alphanumeric), by selecting a cell to make it active and typing the entry from the keyboard.[†] The material you type will be visible in the formula bar as you type. When you have concluded your entry, you press the [Enter↵] key and the entry will appear in the selected cell. By default, the cell below the entry cell then becomes the active one. For our purposes, it would be better for you to change that default. We recommend that you execute the following commands.

Excel commands:

1 Choose **T**ools ➤ **O**ptions...
2 Click on the **Edit** tab
3 Deselect (clear) the **Move selection after Enter** check box

This means after you press [Enter↵] following data entry, you will have to deliberately move to another cell using any of the techniques previously described, like the direction arrow keys or the mouse. We will point out some of the reasons for this as we progress.

In each of the four workbooks that accompany this manual, there is worksheet called **DATA**. As we mentioned, we have entered the data for every example used in the manual. Typically, we have entered the name of the data in the first row of the column and then entered the numerical constants from row 2 on. Thus, for example in the **Descriptive** workbook, there is a set of scores for 12 students on a quiz entered in column **H**. A convenient name is SCORES. So you will find this name entered in cell **H1** as a label; following that you will discover the 12 numerical scores in the cell range **H2:H13**.

[†] To enter a number (such as a phone number) as character data, precede the entry with the apostrophe, '.

Internally, that cell range will be given the name SCORES also for access by functions within Excel.† You do this by selecting cell **H2** and dragging the cell with the mouse to highlight cells through **H13**, the range. With the cells so highlighted, click on the *Name* box (the cell identification will become highlighted) and type in the name SCORES. So, the name in cell **H1** is for your convenience in finding the data set, and the identical name for the range **H2:H13** is for internal use by Excel.

Excel allows for a great deal of flexibility in calling data sets from different worksheets and, indeed, from different workbooks. For example, say you are in the Descriptive workbook and the **DATA** worksheet. You may refer to the Scores data by either the name **SCORES** or its range **H2:H13**. If you are still in the Descriptive workbook but *not* in the **DATA** worksheet, you may still refer to the data by the name **SCORES** but not by the range, since the worksheet you are in has its own column **H**. However, you can still call for the data as **DATA!H2:H13**. Finally, if you are in a different workbook altogether, like Inference, for example, none of the above calls will work. In that case, preceding them with the notation [Descriptive], such as **[Descriptive]DATA!H2:H13**, for instance, will call the proper data set to the Inference workbook.

In many of the Excel routines, it is necessary to capture a range of cells using a range selection box. Such a dialog box (select **Tools** ➤ **Goal Seek**...) is illustrated in Figure 1.6(a) (just ignore the meaning of the tool for now).

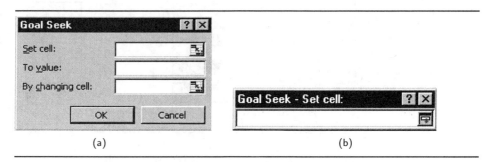

FIGURE 1.6
A range selection dialog box

(a)   (b)

In this dialog box, cell ranges need to be selected for two different text boxes. You can enter the name of a range of data in the text box or the range itself if you remember it. In the right-hand corner of these text boxes you will see a small button, ▦ , called the *collapse dialog button*. When you click this button, the dialog window shrinks to display only the text box you need for entering a range as seen in Figure 1.6(b). This small replacement box is completely movable by dragging its Titlebar with the mouse so that you may view the range you wish to select. When you select that range with the mouse (even in a different worksheet or workbook), the selected cells display a flashing, dotted border (called a *marquee*) and the name of the range is entered in the text box. Clicking ▦ a second time returns you to the original dialog box with that entry accomplished. We will use the phrase: **Use the ▦ button to select**... throughout the rest of this manual to

---

† Some of Excel's functions do not like to find text in the same range with numerical constants.

refer to this technique of clicking the collapse dialog button, making a selection, and then clicking the button a second time to return.

Generally speaking, the data for this manual are already recorded and the specific instructions for you to carry out Excel procedures with the data will be given for each example. You then will be left with at least one companion problem to execute for practice. For some of these, it is simply a matter of replacing the name of the data set previously used with a routine. For others, you will have to mimic the steps provided in the example. In each workbook we have provided a blank worksheet entitled **PRACTICE** where you can and should try data entry and processing on your own.

Many times you have occasion to enter patterned data, that is, data following some sort of sequence or repetition. Excel makes it very easy to do this. Notice that any active cell has a little dark square in its lower right corner. This is called a *fill handle*. When you point to this square, the pointer changes to a filled black cross. To see how this works, choose an empty region in a **PRACTICE** worksheet and type the numbers 2 and 4 in two contiguous rows. Now select those two cells. Point to the fill handle of the cell containing 4, hold the pointer (the black cross) with the mouse and begin to drag it downward. As you drag it, you will see that potential numbers in sequence are suggested for input. See Figure 1.7(a).

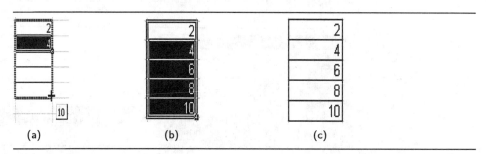

FIGURE 1.7 Patterned data entry

(a)   (b)   (c)

Excel guesses that you meant to put in the sequence 2,4,6,... in successive cells. If you release the mouse at the suggestion of 10, as in Figure 1.7(a), the cells will fill with the suggested values and appear as in Figure 1.7(b). Finally, if you click off the range of cells entirely, you will see the cells as they appear in Figure 1.7(c), ready for processing. The online help facility will give you hints if you are interested in exploring this feature further.

Speaking of processing, another cell entry element is a *formula*, a function that acts on other cell entries be they data or other formulas. You always begin formula entry in Excel by typing the = sign in the formula bar. From that point on there is a large library of formulas available. Some of the simpler ones are obvious from their names. Thus for example suppose the data in Figure 1.7(c) reside in the range **B2:B6**. In cell **B7** (or any other empty cell for that matter), type =SUM(B2:B6). Then the cell containing the formula displays the sum of these cells, 30. Note, however, that the formula itself occupies the cell, not the number 30. So, if you change the entries in the range **B2:B6**, the value will automatically be changed to the sum of the new entries. Incidentally, Excel has a shortcut button for this simple function in the toolbar, ∑, called *AutoSum*. When you click this button, it will place the above formula in the cell that is active at the moment.

Throughout this manual you will be directed explicitly to type in certain formulas; others have already been typed in for you.*(Note:* You should exercise great care in changing the formulas we have provided for you. Always keep an original copy of each workbook at hand in case any of their cells are changed.*)*

MAKING CORRECTIONS

Making corrections in Excel is a relatively simple matter. Just highlight the cell where the error has occured and type in the correct value. The new value appears in the cell as soon as you press the [Enter↵] key. To delete an entry, you merely have to activate the cell and press the [Del] key. If you discover you have made an improper entry, you can "undo" the mistake at that point by selecting the appropriate option from the **Edit** menu.

To insert a cell above the active cell in the Data window choose **Insert** ➤ **Cells**.... You will then be presented with the dialog box shown in Figure 1.8.

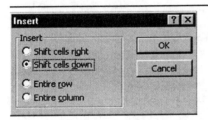

FIGURE 1.8
Inserting cells in a worksheet

You have choices of where the cells should shift after one is inserted or even inserting an entire row or column at this point by clicking the appropriate radio button.

While not strictly under the heading of corrections, the matter of formatting cells is surely one of altering default values. Excel allows many forms of formatting and most of them will be a matter of personal taste. We have used certain ones that we found best for viewing and printing this manual, but you are not bound to them. You will find most of these available alterations in the **Format** menu. First select a range of cells. Then choose **Format** ➤ **Cells**... and you will be presented with the dialog box you see in Figure 1.9 on the following page.

Additional choices within this dialog box may be elected by clicking the different tabs. You can see that the current tab **Number** is open and you see a variety of choices in the **Category** list box, **Number** being one of them. This panel allows you to change the number of decimals places to be viewed[†] in the cell(s), as well as how negative numbers will be viewed.

Other choices in formatting cells include the type of font and font size you might wish to use in the output, what sort of border you would like around the cell(s), and various shades of colors and patterns you might like in the display.

---

† Excel stores all numbers with 15-digit accuracy internally, regardless of what is displayed in the output.

FIGURE 1.9
The Excel Format Cells dialog box

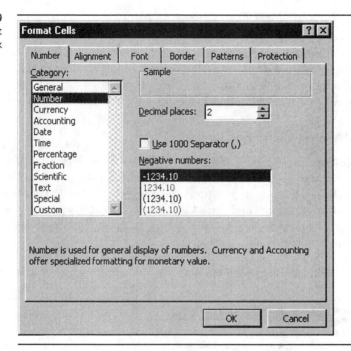

Periodically, we will inform you that we have formatted a display of cells a certain way and usually remind you that you may change that at will without disturbing the contents.

1.4 Excel files

Excel has several different types of files that you may use for permanent storage of results. Our main concern in this manual is with the workbook file, a file ending with the Microsoft extension .XLS. Four such workbooks have been issued to you with this manual. They are designed to address the four main divisions of both *IS/5e* and *ES/4e*. The first, named Descriptive, covers the material found in Parts I and II on sampling and describing data. Part III on probability is covered in the Probability workbook. The Inference workbook covers roughly Part IV of *IS/5e* and most of Part IV of *ES/4e* . Finally, the Regression workbook covers both regression analysis in either book and analysis of variance in *IS/5e* .

OPENING WORKBOOKS

You open a workbook file by choosing **File** ➤ **Open** or clicking 📂 in the toolbar. This will take you to the default location where Excel expects to find workbook files and that location will vary from one configuration to another. Find out where your default location is and copy your manual workbooks (under different names if you like) into that folder. You may change the default location using **Tools ➤ Options...**, then open the General folder and enter a new location in the **Default file location:** text box.

As previously mentioned, each workbook has a **DATA** worksheet and we have taken care to enter the data sets for the various examples and problems of this manual for you. Use these as guides for entering other data sets of your own. The Inference notebook in particular is packed with various formulas used to solve problems in statistical inference. **Alter these at your own risk!** A simple, innocent cell change can affect the entire routine. That is why we advise you to keep the original copy in a safe place and only work with a copy of the original on your computer. Any changes you make will not have any effect on the original copy that way. Should you inadvertently experience problems, you can always save your copy under a different name and re-load the original.

Most of the workbooks also have a **CHARTS** worksheet where you will find any graphics that are generated. We felt it was good housekeeping to separate these outputs from the original data. For one thing we will be using several of the routines provided by Excel in an add-in package found among the analysis tools. You need to be sure that add-in is loaded into your version of Excel. To verify, use the following steps.

Excel commands:

1 Choose **Tools ➤ Add-Ins...**

2 In the Add-Ins dialog box, make sure that the check box for **Analysis ToolPack** has been selected

3 Click OK

Now whenever you select **Tools** from the menu bar, **Data Analysis...** should appear as one of the options. This package, as well as some of the built-in charting routines, are not dynamic. That is to say, you need to run them all over again with a new set of data. This is unlike the formulas inserted in cells of a worksheet where any change in background entry causes an automatic (dynamic) change in the output of the cell.

CLOSING WORKBOOKS

If you exit Excel without any safeguards, all your work will be lost. If you close with the Windows close button, ☒ , a message like that shown in Figure 1.10 appears.

FIGURE 1.10
Exit screen message

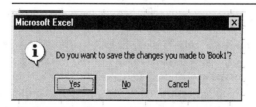

If you select [Yes], you will be presented with a dialog box as shown in Figure 1.11.

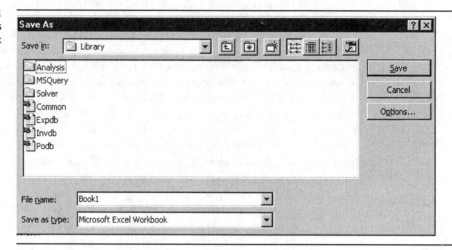

FIGURE 1.11
The Save As dialog box

Of course the entries will vary according to the configuration of your own computer. In the **File name:** text box, type in the name you wish to save this workbook under, say FNAME. This file will then appear in the appropriate folder (your choice) as FNAME.XLS. Later, when you sign on again, you can open this workbook again in a new Excel session using the File menu.

Once your workbook is named, you may always click the save icon in the toolbar, 🖫, at any time to save the current status. Thus, Excel makes losing a worksheet difficult for you by prompting you to consider saving your work. For most of your work here, you will not have to be concerned since all of the material is saved for you on the diskette issued with the manual. However, if you make changes or additions to the material you should save your work and give it a unique name, preserving the original workbooks issued to you intact.

1.5 Simple random sampling

The statistical inference methods studied in an introductory statistics course are designed for use with simple random samples. Given a finite population, a *simple random sampling procedure* is one for which each possible sample of a given size is equally likely to be the one selected. A sample from a population obtained by simple random sampling is often called a *simple random sample*.

Actually, there are two types of simple random samples: namely, those obtained *with replacement* and those obtained *without replacement*. In the former case, sample values may repeat in the sample whereas, no two sample values can be alike in the second sampling procedure. To sample *with replacement*, we use the **Sampling** option in the Data Analysis tool. This will be illustrated in later lessons when we discuss simulation.

Excel has no built-in function to generate samples *without* replacement. That is not a big shortcoming since the bulk of statistical procedures presented in statistics textbooks depends on sampling *with* replacement. The standard procedure, absent a specific function, is to sample with replacement and ignore (or replace) any value that repeats one already generated. Clearly, having to check for ties is cumbersome for large sample sizes. We illustrate with Example 1.11 on page 27 of *IS/5e* and *ES/4e*.

EXAMPLE 1.1 *Illustrates sampling without replacement*

Recall in the example that Professor Hassett wanted to interview a random sample of 15 students from a class of 728 college-algebra students. Since the students were numbered from 1 to 728 on the registration list, he obtained the random sample by selecting 15 numbers from a table of random numbers and matched the numbers to the students to be interviewed. Use Excel to accomplish the same task.

SOLUTION First, store the population of 728 "names" in a column named POPN. The easiest way to enter the integers 1 through 728 into this column is to use Auto fill data entry. Instead of typing each of the integers in the worksheet, Excel allows you to enter a series using only the first and last entry. We proceeded as follows in worksheet CH1 of the Descriptive workbook. (You may repeat the steps in your PRACTICE worksheet if you like.)

Excel commands:

1 Name an empty column POPN (we chose A), type 1 in cell A2, press Enter↵ and be sure A2 is selected
2 Choose Edit ➤ Fill ➤ Series...
3 Select the Columns option button in the Series in option box
4 Select the Stop value: text box and type 728
5 Click OK

The successive integers from 1 through 728 are now stored in POPN as you may verify. *(Note:* For small populations, see page 10 for instructions on using the fill handle.) Name an empty column SMPL (we chose B). Then create and store the random sample in the column named SMPL as follows.

Excel commands:

1 Choose Tools ➤ Data Analysis...
2 Select Sampling in the Analysis Tools option box
3 Type in the range of data for POPN, (A2:A729 in our case) in the Input Range: text box
4 Click the Random radio button in the Sampling Method option box, select the Number of Samples: text box and type 15

5 Select the **Output Range:** option button in the **Output options** option box, select its text box and type B2

6 Click [OK]

7 Click the sort button, [A↓Z], in the Excel toolbar

Now the registration numbers for 15 students will be stored in SMPL in sorted order from low to high. You may view the result by examining the worksheet. See Printout 1.1 for our result, which will probably differ from yours.

PRINTOUT 1.1
Excel SMPL output

68	89	158	158	226	267	299	315
356	370	372	383	396	493	617	

As it happens, cells B4 and B5 have the same value. To break this tie, click on either cell, say B5, and repeat the Sampling procedure replacing 15 by 1 in the **Number of Samples:** text box (Step 4), and replace B2 by B5 in the **Output Range:** text box (Step 5). When asked, accept the overwrite of cell B5 by clicking [OK]. See Printout 1.2 for the result.

PRINTOUT 1.2
Revised Excel SMPL output

68	89	158	78	226	267	299	315
356	370	372	383	396	493	617	

Notice that 158 has been replaced by 78 in the revised sample. (Of course another attempt would replace it with something other than 78 in all likelihood.) The students whose registration numbers correspond to those in this revised list would then be interviewed. ∎

PROBLEM 1.1 Refer to Example 2.2. Use Excel to obtain your own random sample of 15 students to be interviewed.† ◊

1.6 Systematic random sampling [E-34:I-34]

Example 1.13 on page 33 of *IS/5e* and *ES/4e* illustrates the procedure for constructing a systematic random sample from the population of 728 students treated in the last example. The only time that the random number generator is used in that technique is to select the first number to be included in the sample. This is accomplished as follows.

---

† The solutions to all problems can be found in Appendix A. Incidentally, problems end with the symbol ◊; the symbol ∎ is used to signal the end of an example.

1.6 Systematic random sampling

Let N be the population size (728 in the example), and let n be the sample size (15 in the example). Find the factor K where $K = N/n$ rounded down (48 in the example). Treating the integers 1 through K as a new population, draw a sample of size 1 from this population and treat this as the first sample value. The remaining values are obtained by adding K to each successive value as in the next example.

EXAMPLE 1.2 *Illustrates the creation of systematic samples in Excel*

To see an immediate application, take the same circumstances as those of Example 1.13 on page 33 in *IS/5e* and *ES/4e*. In the **CH1** worksheet, name three successive columns **SYSTF**, **SYSTP**, and **SYSTS** (we chose columns C, D, and E). Proceed as follows.

Excel commands:

1. Select cell **C2** in column **SYSTF** and type `=ROUND(728/15-0.5,0)`
2. Select cell **D2** in column **SYSTP**, type 1, press Enter⏎ and be sure **D2** is selected
3. Choose **E**dit ➤ F**i**ll ➤ **S**eries...
4. Select the **C**olumns option button in the **Series in** option box
5. Select the **St**op value: text box and type 48
6. Click [OK]

Now we are ready to create the systematic sample values in column **SYSTS**. Proceed as follows.

Excel commands:

1. Choose **T**ools ➤ **D**ata Analysis...
2. Select **Sampling** in the **A**nalysis Tools option box
3. Type in the range of data for SYSTP (D2:D49 in our case) in the **I**nput Range: text box
4. Select the **R**andom radio button in the **Sampling Method** option box, select the **Number of Samples:** text box and type 1
5. Select the **O**utput Range: option button in the **Output options** option box, select its text box, type E2 and click [OK]
6. Select cell **E3** and type `=E2+$C$2`
7. Drag the Fill Handle for **E3** through cell **E16**

That's it. The systematic sample is now located in column SYST. Our results are displayed in Printout 1.3.

PRINTOUT 1.3
Excel systematic sample

```
 23   71  119  167  215  263  311  359
407  455  503  551  599  647  695
```

As you see, this is similar to the sample displayed in Table 1.11 on page 34 of *IS/5e* and *ES/4e*. It happens that the initial value in this case is 23 instead of 22; otherwise, the systematic sample is constructed just as required. The initial value was selected at random, and the difference between successive values in the sample is exactly 48. Since there were no ties to break, the sample stands as displayed.

It should be noted that some items in the population may never be selected in a systematic sample. For example, the 8 students numbered 721 through 728 can never appear in a systematic sample in this example. Indeed, the largest value that will ever occur is 720 since the largest initial value possible in the sampling scheme is 48 and $15 * 48 = 720$. ∎

PROBLEM 1.2   Refer to Example 1.2. Use Excel to obtain your own systematic random sample of 15 students to be interviewed.   ◇

LESSON 2

Organizing Data

GENERAL OBJECTIVE In Lesson 1 you learned some basic commands that are necessary for using Excel. You also learned how to generate samples to form data sets. Now you will see how Excel can be employed to organize data into tables, and summarize data with graphical displays.

LESSON OUTLINE
2.1 **Grouping data**
2.2 **Graphs and charts**

2.1 Grouping data

E-70 : I-70

We often gain insight into a set of data using various grouping methods. By grouping the data into classes, we may tally the values in each class and often calculate relative frequencies to obtain a frequency distribution. We call this a *grouped-data table*. Excel allows us to form a simple frequency table through its **Histogram** (see the next section for more) routine in the **Data Analysis** tool. From this output we can fill in the rest of the table with appropriate Excel functions. We show how this is accomplished in the next example.

EXAMPLE 2.1   *Illustrates grouped-data tables*

The U.S. National Center for Health Statistics publishes data on weights and heights by age and sex in *Vital and Health Statistics*. The weights in Table 2.1, given to the nearest tenth of a pound, were obtained from a sample of 18–24-year-old males. Construct a grouped-data table for these weights. Use a class width of 20 and a first cutpoint of 120.

TABLE 2.1
Weights of 37 males, age 18-24 years

129.2	185.3	218.1	182.5	142.8	155.2	170.0	151.3
187.5	145.6	167.3	161.0	178.7	165.0	172.5	191.1
150.7	187.0	173.7	178.2	161.7	170.1	165.8	
214.6	136.7	278.8	175.6	188.7	132.1	158.5	
146.4	209.1	175.4	182.0	173.6	149.9	158.6	

SOLUTION   For problems in this lesson, the data sets are stored and named in the **DATA** worksheet of the **Descriptive** workbook and the solutions are presented in the **CH2** worksheet. Click the tab for the **DATA** worksheet and you will find the data from Table 2.1 stored under the name WEIGHTS.

Click the tab for the **CH2** worksheet. In column **A**, you will find the shaded title **DATA FILE NAME** with a dotted border around cell **A3**. This is our way of denoting an input cell, a cell that requires something from you, the reader. This will always be true in the workbooks we have provided you. In this case, we need to input the filename WEIGHTS and, since that file is located in the same worksheet, it is not necessary to identify its location as part of the **DATA** worksheet. Also, it will not matter whether or not the input is in lower case or upper case letters, that is, the input cell is not *case sensitive*. Continue as follows.

Excel commands:

1   Select input cell **A3**, type WEIGHTS and press Enter↵

In the output range **B1:C5** depicted in Figure 2.1 on the following page, you will find a summary of numbers that will help you determine the classes for the grouped data.

In **C1** you see that the count, or sample size, of the data is 37. Then cells **C2** and **C3** automatically compute the maximum and minimum values in the data set. Next, you will find the class width that would be required to have 5 classes altogether (cell **C4**) and the width that would be required to have 20 classes altogether (cell **C5**). These are only suggestions and may be overridden. However, you must make a decision and input your choice of a width in input cell **C6**. We will use 20 for this illustration and you may experiment with other choices.

FIGURE 2.1
Grouped-data table for WEIGHTS data in Excel

	A	B	C	D	E	F	G	H
1	DATA FILE	COUNT =>	37		Bin	Frequency	Rel. Freq.	Midpoint
2	NAME :	Max =>	278.8		140.0	3	0.081	130.0
3	WEIGHTS	Min =>	129.2		160.0	9	0.243	150.0
4		Width(5) =>	29.9		180.0	14	0.378	170.0
5		Width(20) =>	7.5		200.0	7	0.189	190.0
6		Width ==>	20		220.0	3	0.081	210.0
7		Cutpoints =	Bin		240.0	0	0.000	230.0
8		First L Cutpt=>	120.0		260.0	0	0.000	250.0
9		First U Cutpt=>	140.0		280.0	1	0.027	270.0

In row 7 you will see the term *Bin* equated to *Cutpoints*. Excel considers the classes of a grouped table to be like a bin into which you are placing certain data values. If you know the width, you really only need to know the upper (or lower) cutpoint of each bin. So Excel simply lists the Upper cutpoint of each class, calling that a Bin, but it stands for the entire class from that point down one class width.

However, in contrast to your book, Excel adopts the convention of including an item in the class if it is *greater* than the lower cutpoint and *less than or equal* to the upper cutpoint. Thus, an entry exactly at the upper cutpoint value would always belong to the class below that value and not above the cutpoint as in your book. We can adjust to accommodate this conventional difference by judicious choices of cutpoint values as follows.

Excel commands:

1. Select input cell **C6**, type 20 and press [Enter ↵]
2. Select **C8**, type 119.99 and press [Enter ↵]
3. Select **C9**, type 139.99 and press [Enter ↵]
4. Select the range **C8:C9** and drag the fill handle of cell **C9** down through enough cells to cover the maximum value of 278.8 (that would be cell **C16** for this example)

(*Note:* By formatting the cells in column **C** we force the output to display only one decimal place even though the entries consist of 2 decimal places.) We completed construction of the grouped data table shown in Figure 2.1 by executing the Excel commands on the following page.

22 Lesson 2 Organizing Data

Excel commands:

1. Choose **T**ools ➤ **D**ata Analysis... and select **Histogram** from the **A**nalysis Tools menu
2. Type `WEIGHTS` in the **I**nput Range: text box
3. Use the ▨ button to select a range for the **B**in Range: text box (that would be **C9:C16** for this example)
4. Select the **O**utput Range: radio button in the Output Options box, type **E1** in the text box, and deselect all check boxes
5. Click [OK]

Figure 2.1 displays the output in the range **E1:H9**. Compare this output to Table 2.6 on page 71 of *IS/5e* and *ES/4e*, a grouped-data table found by hand. Actually, the output from the Data Analysis tool only consists of columns E and F. We have added columns G and H to emulate the output in Table 2.6 of *IS/5e* and *ES/4e*.[†]

For a new set of data included in the **DATA** worksheet, you have only to overwrite the name in the input cell **A3**. (See page 9 if your data set is located in a different workbook.) Then execute the above Excel commands. You also need to consider formatting the output cells so that the entries reflect the decimal places in the data and midpoints are given with no more than 1 decimal place beyond the data, a convention we try to follow in this manual.

PROBLEM 2.1   The following table gives the number of days to maturity for 40 short-term investments. The data are from *Barron's National Business and Financial Weekly*.

70	64	99	55	64	89	87	65
62	38	67	70	60	69	78	39
75	56	71	51	99	68	95	86
57	53	47	50	55	81	80	98
51	36	63	66	85	79	83	70

We have stored these data under the name MATURITY in the **DATA** worksheet of the **Descriptive** workbook. Construct a grouped-data table, using Excel. Use $30 \leqslant 40$ for the first class.   ◊

SINGLE-VALUE GROUPING   E-73 : I-73

For discrete data it is often preferable to use each single value as a "class." In that case we only have to be certain that the single value is the midpoint of class as we show in the next example.

[†] Excel always adds an extra row labeled **More** with a frequency of 0 to indicate that there are no data values above the last upper cutpoint. We do not display that row in Figure 2.1. If you find a nonzero entry there, you have not selected the proper range for the Excel **B**in range: text box in Step 3 above.

EXAMPLE 2.2 *Illustrates single-value grouping*

A planner is collecting data on the number of school-age children per family in a small town. Thirty families are selected at random. The following table displays the number of school-age children in each of the families chosen.

TABLE 2.2
Number of school-age children in each of 30 families

0	3	0	0	3	0
2	2	0	1	2	1
0	0	1	2	4	0
4	2	1	0	1	0
0	2	0	1	3	2

Use Excel to obtain a grouped-data table for these data using classes based on a single value.

SOLUTION The data have been entered in the **DATA** worksheet of the Descriptive workbook under the name **CHILDREN**. Open the **CH2** worksheet and proceed as follows. We want the cell width to be 0 so that upper and lower cutpoints agree (hence you may ignore input cell **C8**). Figure 2.2 shows the output resulting from executing the Excel commands following the figure.

FIGURE 2.2
Grouped-data table for single-value grouping in Excel

Bin	Frequency	Rel. Freq.	Midpoint
0	12	0.400	0.0
1	6	0.200	1.0
2	7	0.233	2.0
3	3	0.100	3.0
4	2	0.067	4.0

Excel commands:

1 Type <u>CHILDREN</u> in the input cell **A3**
2 Type <u>0</u> in the input cell **C6**
3 Select cell **C9**, type <u>0</u> and press [Enter ↵]
4 Select cell **C10**, type <u>1</u> and press [Enter ↵]
5 Select the range **C9:C10** and drag the fill handle of cell **C10** down through enough cells to cover the maximum value of 4 (that would be cell **C13** for this example)
6 Choose **Tools ▶ Data Analysis...** and select **Histogram** from the **Analysis Tools** menu
7 Type <u>CHILDREN</u> in the **Input Range:** text box
8 Use the ▦ button to select a range for the **Bin Range:** text box (that would be **C9:C13** for this example)
9 Select the **Output Range:** radio button in the Output Options box, type <u>E1</u>, deselect all check boxes, and click [OK]

24   Lesson 2   Organizing Data

The output is consistent with the fact that each numerical value (under **Bin**) constitutes a class. Compare these results with Table 2.9 on page 72 of *IS/5e* and *ES/4e*. ∎

PROBLEM 2.2   The U. S. Bureau of the Census conducts nationwide surveys on characteristics of American households. Below are data on the number of persons per household for a sample of 40 households.

2	5	2	1	1	2	3	4
1	4	4	2	1	4	3	3
7	1	2	2	3	4	2	2
6	5	2	5	1	3	2	5
2	1	3	3	2	2	3	3

Construct a grouped-data table based on single-value grouping using Excel. The data are stored in the **DATA** worksheet under the name **PERSONS**.   ◇

2.2   Graphs and charts   ⬚ E-82 : I-82

Excel has commands that will simultaneously group a data set and display the result in graph called a frequency *histogram*. We will now illustrate using the days-to-maturity data that we considered in Problem 2.1.

EXAMPLE 2.3   *Illustrates constructing a histogram with Excel*

The number of days to maturity for 40 short-term investments, given in Problem 2.1, is repeated here as Table 2.3.

TABLE 2.3
Days to maturity for 40 short-term investments

70	64	99	55	64	89	87	65
62	38	67	70	60	69	78	39
75	56	71	51	99	68	95	86
57	53	47	50	55	81	80	98
51	36	63	66	85	79	83	70

Use Excel to construct a frequency distribution and frequency histogram for these data based on the classes $30 < 40, 40 < 50, \ldots, 90 < 100$.

As you might have guessed, the **Histogram** Analysis Tool has been designed to carry out this task. The data are under the name **MATURITY**. It is as simple as grouping the data as in the last section (adapting the Excel commands on page 23) and this time selecting the **Chart Output** check box in the **Output Options** options box in addition to the **Output Range:** option at Step 9. (We chose **J2** for the output cell.) You should see a chart that resembles the one in Figure 2.3 on the following page.

FIGURE 2.3
Excel's histogram for MATURITY data

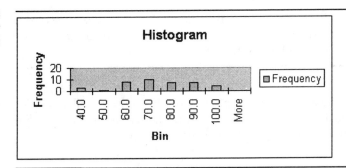

Of course this image is unfavorable from many points of view. For one thing it is too small to create a good impression. And, the rectangles that make up the histogram do not fill in the interval scale below it, as in Figure 2.1 on page 82 of *IS/5e* and *ES/4e*. We modified the graph to appear as in Figure 2.4 executing the Excel commands below the figure. You may select your own modifications of course.

FIGURE 2.4
Excel's histogram for MATURITY data

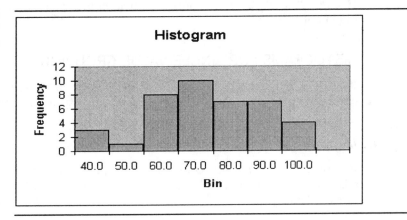

Excel commands:

1. Delete the category **More** in the grouped table output
2. Click anywhere in the figure
3. Delete the **Frequency** box to the right of the figure
4. Drag the middle button of the bottom line down to cover around 15 rows
5. Right click any rectangle in the histogram and Select **F**ormat Data Series...
6. Select the **Options** tab and make the **G**ap width 0 by holding the down button next to its text box until 0 appears (or type **0** in the text box)
7. Click [OK]

This is an improvement over Figure 2.3 and resembles Figure 2.1(a) on page 82 of *IS/5e* and *ES/4e* more closely. Also, each upper cutpoint covers or stands for a specified interval or bin as it should. For example, place the mouse pointer anywhere in the rectangle over upper cutpoint 60.0, and you will be notified that the interval, $50 < 60$, the bin, contains 8 observations. ∎

PROBLEM 2.3 The U.S. Bureau of Labor Statistics collects and publishes data on the ages of persons, 16 years old and over, in the civilian labor force. Suppose that the ages of a sample of 50 such persons are as follows:

22	58	40	42	43	32	34	45	38	19
33	16	49	29	30	43	37	19	21	62
60	41	28	35	37	51	37	65	57	26
27	31	33	24	34	28	39	43	26	38
42	40	31	34	38	35	29	33	32	33

Use Excel to obtain a frequency histogram of the data. Take the classes to be $15 < 20$, $20 < 25$, $25 < 30$, and so on. The data list is named **AGES** in the **DATA** worksheet. ◊

HISTOGRAMS FOR SINGLE-VALUE GROUPING `E-84 : I-84`

If we want our frequency distribution and histogram to be based on single-value classes, then we simply use a class width of 1 in setting the interval parameters.

EXAMPLE 2.4 *Illustrates a histogram for single-value grouping*

Return to the circumstances of Example 2.2 on page 23 dealing with the number of school-age children per family in a small town. Create a histogram for the grouping formed there.

SOLUTION Repeat the steps of Example 2.3 to produce the result in Figure 2.5.

FIGURE 2.5
Excel's histogram for CHILDREN data

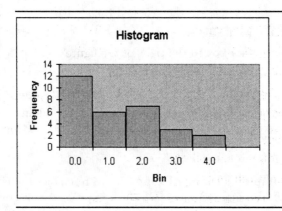

As mentioned earlier, lower and upper cutpoints agree with the single values themselves in this case. Still, the cutpoint reflects the frequency in the bin it represents, namely, the single value. ∎

PROBLEM 2.4 The U.S. Department of Housing and Urban Development (HUD) compiles data on the number of rooms in occupied housing units. Suppose that a sample of 25 housing units yields the following data on number of rooms.

4	4	6	7	7
8	4	4	8	5
2	9	3	6	7
5	3	5	8	5
6	4	6	5	5

Employ Excel to construct a frequency distribution and frequency histogram of the data using classes based on a single value. The data are named **ROOMS** in the **DATA** worksheet. ◊

DOTPLOTS

E-85 : I-85

On page 84 of *IS/5e* and *ES/4e*, dotplots are discussed. Recall that a *dotplot* for a data set is constructed by displaying the possible numerical values on a horizontal axis and by recording each piece of data with a dot over the appropriate value on the horizontal axis. This might be called a "poor man's" histogram for it conveys about the same information. Excel has no chart option for creating dotplots, but its histogram capability can be used to advantage in such cases.

EXAMPLE 2.5 *Illustrates an Excel version of a dotplot*

A farmer is interested in a new fertilizer that supposedly will increase his yield of oats. He uses the fertilizer on a sample of 15 one-acre plots. The yields, in bushels, are depicted in Table 2.4. Use Excel to construct a dotplot of the data.

TABLE 2.4
Oat yields

67	65	55	57	58
61	61	61	64	62
62	60	62	60	67

SOLUTION The data are stored in a column named **OATS** in the **DATA** worksheet. Create a histogram for these data as a single-valued data set as follows.

Excel commands:

1 Type <u>OATS</u> in the input cell **A3**
2 Type <u>0</u> in input cell **C6**
3 Select cell **C9**, type <u>55</u> and press ⏎

28 Lesson 2 Organizing Data

4 Select cell **C10**, type <u>56</u> and press [Enter ↵]
5 Select the range **C9:C10** and drag the fill handle of cell **C10** down through enough cells to cover the maximum value of 67 (that would be cell **C21** for this example)
6 Choose **Tools ➤ Data Analysis...** and select **Histogram** from the **Analysis Tools** menu
7 Type <u>OATS</u> in the **Input Range:** text box
8 Select the range **C9:C21** for the **Bin Range:** text box
9 Select any output range that is clear on your workspace, and be sure that the **Chart Output** check box is selected
10 Click [OK]

That much will produce a basic histogram for the OATS data and we ask you to modify the chart as follows.

Excel commands:

1 Delete the word **More** in the table output
2 Click anywhere in the chart output to select it then stretch it to cover about 10 rows
3 Delete the legend to the right of the histogram
4 Format the **Category Axis** as a number with 0 decimal places
5 Click on the **Category Axis Title**, type <u>OATS</u> and press [Enter ↵]
6 Click on the **Chart Title** at the top of the histogram and type `Bar Chart`
7 Right click any bar in the histogram, select **Format Data Series...**, click the **Options** tab, type <u>500</u> in the **Gap width:** text box
8 Click [OK]

Figure 2.6 displays the resulting output after this formatting.

FIGURE 2.6
Bar chart of OATS data using Excel

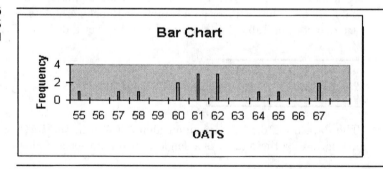

You should compare the bar chart to the dotplot in Figure 2.3 on page 85 of *IS/5e* and *ES/4e*. It surely conveys the same information. ∎

PROBLEM 2.5 The following table records the scores for a quiz in a beginning statistics course.

60	51	57	42
51	43	43	55
52	51	50	50

The data are stored in a column named SCORES in the **DATA** worksheet. Use Excel to construct a bar chart of the data. ◊

LESSON 3

Descriptive Measures

GENERAL OBJECTIVE Besides organizing data into tables and portraying data with graphs, we can also summarize data using descriptive measures. Recall that *descriptive measures* are numbers used to describe data sets. In this lesson we will learn how Excel can be employed to compute various descriptive measures.

LESSON OUTLINE
3.1 Measures of center
3.2 The sample mean
3.3 Measures of variation; the sample standard deviation
3.4 The five-number summary
3.5 Descriptive measures for populations

3.1 Measures of center

Descriptive measures that indicate where the center or most typical value of a data set lies are called *measures of central tendency* or, more simply, *measures of center*. These are also referred to by many as *averages*. In this section we will see how Excel can be used to obtain the two most important measures of central tendency, the *mean* and the *median*.

THE MEAN

`E-127 : I-127`

Recall that the *mean* of a data set is defined to be the sum of the data divided by the number of observations. The Excel function for computing the mean of a data set is called **AVERAGE**.

EXAMPLE 3.1 *Illustrates the AVERAGE function*

A mathematician spent a summer working for a small mathematical consulting firm. The firm employed a few senior consultants who made between $800 and $1050 per week, a few junior consultants who made between $400 and $450 per week, and several clerical helpers who made $300 per week. There was more work in the first half of the summer than in the second half, so there were more employees during the first half. Tables 3.1 and 3.2 give typical lists of weekly earnings for the two halves of the summer. Use Excel to find the mean of each of the two sets of data.

TABLE 3.1 Data Set I

$300	300	300	940	300	300	400
300	400	450	800	450	1050	

TABLE 3.2 Data Set II

$300	300	940	450	400
400	300	300	1050	300

SOLUTION These two sets of data are located in the worksheet **DATA** of the **Descriptive** workbook under the names **DATAI** and **DATAII**. Press the **CH3** worksheet tab and proceed as follows.

Excel commands:

1 Select any empty cell and format it for currency with 2 decimal places
2 Type `=AVERAGE(DATAI)`

The value $483.85 will appear in the cell instantly. Repeat this exercise by replacing **DATAI** with **DATAII** (or their respective cell references in the **DATA** worksheet) and you will see $474.00 appear in the cell. Thus the mean salary of the employees in Data Set I is $483.85 and that of the employees in Data Set II is $474.00. ∎

PROBLEM 3.1 A college chemistry teacher is concerned about the possible detrimental effects of a poor mathematics background on his students. He takes a sample of 15 students and divides them according to their background in mathematics. Their semester averages turn out to be the following:

Fewer than two years of high school algebra	Two or more years of high school algebra
58	84
81	67
74	65
61	75
64	74
43	92
	83
	52
	81

Use Excel to determine the mean of each of the two data sets given in the table. In the **DATA** worksheet, LTH2 is the name for the data in the first column and GTH2 is the name for the other column of data. (You may need to format your cell(s) accordingly.) ◊

THE MEDIAN   E-128 : I-128

Recall that the *median* of a data set is defined to be (1) the data value exactly in the middle of its ordered list, if the number of pieces of data is *odd*, or (2) the average of the two middle data values in its ordered list, if the number of pieces of data is *even*. As you might suspect, the appropriate Excel function is called MEDIAN.

EXAMPLE 3.2 *Illustrates the MEDIAN function*

Consider again the two sets of salary data in Tables 3.1 and 3.2 on page 31. Employ Excel to find the median of each of the two data sets.

SOLUTION We have only to repeat the last exercise, replacing the name AVERAGE with MEDIAN. You may easily verify that the median salary of the employees in Data Set I is $400 and that of the employees in Data Set II is $350. ∎

PROBLEM 3.2 Refer to Problem 3.1. Use Excel to find the median of the two data sets. ◊

THE MODE   E-129 : I-129

Excel also has a MODE function that may be used in the same manner as the last two we discussed.

EXAMPLE 3.3 *Illustrates finding the Mode*

Consider again the two sets of salary data given in Tables 3.1 and 3.2 on page 31. Repeat the last example replacing the function MEDIAN by the function MODE. By inspection verify that the mode for DataI is $300. Also, DataII has the same unique mode.

PROBLEM 3.3 The U.S. Department of Housing and Urban Development (HUD) compiles data on the number of rooms in occupied housing units. Suppose that a sample of 25 housing units yields the following data on number of rooms:

4	4	6	7	7
8	4	4	8	5
2	9	3	6	7
5	3	5	8	5
6	4	6	5	5

Use Excel to find the mode of the number of rooms. The data set is named HUD in the DATA worksheet. ◊

3.2 The sample mean

The sample mean is just the mean of a set of sample data, which is the sum of the observations divided by n, the sample size. Both of these numbers can be computed directly using Excel calculator functions.

SUMMATION   E-138 : I-138

Frequently we need to find the sum of a data set. By employing Excel, we can easily obtain the sum of any data set. The appropriate function is called SUM.

EXAMPLE 3.4 *Illustrates the SUM function*

A student takes four exams in a biology class. His grades are 88, 75, 95, and 100. Find the sum of this grade data with the aid of Excel.

SOLUTION First, enter the data into any column in the CH3 worksheet. Then, in the cell following the last data value, click the Σ button (called AutoSum) in the tool bar, and you will see the sum of 358 appear in that cell instantly. Highlight the cell containing the sum again and look in the formula bar. You should see =SUM(A1:A4) (range depending on your placement of the data) displaying the SUM function with its argument, a range or the name of a range. Typing this in the formula bar is an

34 Lesson 3 Descriptive Measures

alternative to clicking the Σ button. You can have as many as 30 such arguments separated by commas. ∎

PROBLEM 3.4 The following table gives the numbers of patients admitted to U.S. hospitals between 1977 and 1982. [Source: American Hospital Association.]

Year	Patients admitted (millions)
1977	37.1
1978	37.2
1979	37.8
1980	38.9
1981	39.2
1982	39.1

Use Excel to determine the total number of patients admitted to U.S. hospitals between 1977 and 1982, inclusive. We have used PATIENTS for a column name. ◊

THE SAMPLE MEAN   E-140 : I-140

The sample mean is just the mean of a set of sample data, which is the sum divided by n. The value of n may be found in Excel using the COUNT function. Thus, once data are stored in a column, the value of $\bar{x}$ may be computed using Excel functions.

EXAMPLE 3.5 *Illustrates the computation of the sample mean*

Each year manufacturers perform mileage tests on new car models and submit the results to the Environmental Protection Agency (EPA). The EPA then tests the vehicles to determine whether the manufacturers are correct.

In 1998, one company reported that a particular model, equipped with a four-speed manual transmission, averaged 29 miles per gallon (mpg) on the highway. Let us suppose that the EPA tested 15 of the cars and obtained the gas mileages given in Table 3.3.

TABLE 3.3
Gas mileages

27.3	31.2	29.4	31.6	28.6
30.9	29.7	28.5	27.8	27.3
25.9	28.8	28.9	27.8	27.6

Determine the sample mean of these gas mileages using Excel functions.

SOLUTION We have entered the data in a column named EPA. We need to find the sum of the data and divide by their number. Finally, we need to divide these two quantities to compute the sample mean.

3.3 Measures of variation; the sample standard deviation

Excel commands:

1. Select any empty cell and format it for a number with 1 decimal place
2. Type =SUM(EPA)/COUNT(EPA) in the formula bar and press ⏎

You should see the value 28.8 in the cell instantly. Of course, to find the sample mean using Excel, we need only apply the AVERAGE function. As you may verify, the results agree. ∎

PROBLEM 3.5 The U.S. Federal Bureau of Investigation (FBI) conducts annual surveys to estimate the average value lost by victims of selected crimes. Suppose that a sample of shoplifting offenses yields the following data on value lost.

$88	70	121	159
41	39	82	107
69	53	18	95

The data are located in a column named FBI. Determine the sample mean of these losses with the aid of Excel using the functions SUM and COUNT. Compare your answer with the result of applying the AVERAGE function. ◇

3.3 Measures of variation; the sample standard deviation

Descriptive measures that indicate the amount of variation in a data set are called *measures of variation* or *measures of spread*. We will learn in this section how Excel can be applied to obtain two of the most frequently used measures of spread, the *range* and the *standard deviation*.

THE RANGE   E-143 : I-143

The *range* of a data set is the difference between the largest and smallest data values in the data set:

Range = Largest value − Smallest value.

Excel has a function MAX that will compute the maximum value and another one MIN that will compute the minimum value of a set of data. From these two outputs, it is a simple matter to compute the range.

EXAMPLE 3.6 *Illustrates how to find the range using Excel*

The heights, in inches, of the five starting players on a basketball team are 67, 72, 76, 76, and 84. Use Excel to obtain the range of these height data.

SOLUTION The height data are stored in a column named HEIGHT. So, select any empty cell and type the formula =MAX(HEIGHT)-MIN(HEIGHT). The value of the range will appear instantly as 17 inches. ∎

PROBLEM 3.6 Repeat Example 3.7 for the data of Problem 3.5 on page 35. Apply the appropriate Excel functions to find the range of the losses. ◊

THE SAMPLE STANDARD DEVIATION `E-150:I-150`

Recall that the *sample standard deviation*, *s*, of *n* pieces of sample data is defined by

$$s = \sqrt{\frac{\Sigma(x-\bar{x})^2}{n-1}}$$

and can also be computed by using the shortcut formula

$$s = \sqrt{\frac{\Sigma x^2 - (\Sigma x)^2/n}{n-1}}.$$

This last formula is quite suitable for use by calculating devices that are capable of storing three quantities from the data. These three quantities are the sample size, n; the sum, Σx; and the sum of squares, Σx^2. Most of the hand-held calculators on the market today have this capability.

Regardless of which formula is used to compute s, the calculations are generally tedious and time-consuming. We can avoid doing the calculations by employing Excel's STDEV function.

EXAMPLE 3.7 *Illustrates the STDEV function*

Consider again the heights, in inches, of the five starting players on a team as discussed in Example 3.7. The heights are 67, 72, 76, 76, and 84. Use Excel to determine the sample standard deviation of these heights.

SOLUTION In Example 3.7 we stored the heights in a column designated HEIGHT. To obtain the sample standard deviation of those heights, we proceed as follows. Locate an empty cell and type =STDEV(HEIGHT). The cell contents should display the value 6.2450 instantly. ∎

PROBLEM 3.7 Determine the sample standard deviation of the loss data given in Problem 3.5 on page 35 with the aid of Excel. ◊

3.4 The five-number summary

Several data measures occur in applications repeatedly and it is always nice to be able to discover them quickly. It is common for every statistical package to have routines for computing these summaries. Excel has an option in the **Data Analysis** tool called **Descriptive Statistics** that will display several descriptive measures of a data set simultaneously. But first, we examine the problem of computing quartiles with Excel.

QUARTILES; INTERQUARTILE RANGE $\quad\cdot\quad$ E-162 : I-162

Excel has a function called **QUARTILE** that will compute quartiles directly. The *Note* on page 160 of *IS/5e* and *ES/4e* mentions that not even all statisticians are in agreement of the definition of quartiles so it is not surprising that various definitions are used by different computer software packages. Excel uses a definition different than that given in Definition 3.7 on page 160 of *IS/5e* and *ES/4e*. We will discuss both in this manual. We will illustrate with the following example.

EXAMPLE 3.8 *Illustrates the use of the QUARTILE function*

A sample of 20 people yields the weekly television viewing times, in hours, displayed in Table 3.4.

TABLE 3.4 Weekly TV viewing times

25	41	27	32	43
66	35	31	15	5
34	26	32	38	16
30	38	30	20	21

Use Excel to locate the first and third quartiles for this data set.

SOLUTION We have already stored the data in the **DATA** worksheet under the name **TIMES**. The Excel function QUARTILE takes two arguments, the range of data and the quartile of interest. Choose an empty cell in the Descriptive worksheet and type `=QUARTILE(TIMES,1)`. You will find the value 24 displayed in the cell instantly. Similarly, to find the third quartile, type `=QUARTILE(TIMES,3)`. You will find the value 35.75 displayed in the cell instantly. Thus the interquartile range is $IQR = 35.75 - 24 = 11.75$. These values differ from those derived in your textbook because a different definition of quartile is used by Excel.

We can still use Excel to generate the necessary formulas to accommodate Definition 3.7 of *IS/5e* and *ES/4e*. The first thing we need to do is sort the data as is done in the book. This is easy. In the **DATA** worksheet copy the data from **TIMES** to a new column called **STIMES** as we have† and, while selected, click the

---

† using the copy and paste buttons in the toolbar.

sort button, [A↓Z], in the Excel toolbar. The data are now arranged from low to high. We proceed with the steps carried out in *IS/5e* and *ES/4e*.

We first need to calculate the quantity $(n+1)/4$ and then attempt to locate a point on the scale as "deep" as this amount. This is the first quartile, Q_1. Similarly, the third quartile is located at a depth of $3(n+1)/4$ in the data set. Since $n=20$, $(n+1)/4 = 5.25$. From the ordered data we need to find a depth of 5.25, and this takes us to a point between 21 and 25. One fourth of the distance of 4 between these points is 1, so $Q_1 = 22.0$.

Similarly, $3(n+1)/4 = 15.75$. Since the 15th data value is 35 and 16th data value is 38, a depth of 15.75 takes us three fourths of the distance of 3 from 35 to the point $35 + 2.25 = 37.25$. This is the value of Q_3.

Using the definition of interquartile range on page 162 of *IS/5e* and *ES/4e*, $IQR = Q_3 - Q_1 = 37.25 - 22 = 15.25$. As already noted, Q_1, hence IQR, are both different from those obtained from the Excel **QUARTILE** command. ∎

PROBLEM 3.8   The U.S. Energy Information Administration publishes figures on residential energy consumption and expenditures. Suppose that a sample of 36 households using electricity as their primary energy source yields the following data on last year's energy expenditures.

$1376	1452	1235	1480	1185	1327
1059	1400	1227	1102	1168	1070
1180	1221	1351	1014	1461	1102
976	1394	1379	987	1002	1532
1450	1177	1150	1352	1266	1109
949	1351	1259	1179	1393	1456

The data have been stored in a column named **ENERGY**. Find the first and third quartiles using Excel's function. Use these values to compute the value of the interquartile range. Compare these results with those obtained from Definition 3.7 of *IS/5e* and *ES/4e*. ◇

THE FIVE-NUMBER SUMMARY   [E-163 : I-163]

The **Descriptive Statistics** option among the **Data Analysis** tools in Excel function will calculate and display a wide variety of summaries that might be of interest.

EXAMPLE 3.9   *Illustrates the use of the Descriptive Statistics Data Analysis tool*

Use the **Descriptive Statistics Data Analysis** tool in Excel to obtain several descriptive measures of the television viewing time data given in Table 3.4 of Example 3.8 on page 37 of this manual.

SOLUTION   The data have been stored for you in a range named **TIMES** in the **DATA** worksheet. Click on the **CH3** worksheet tab and proceed as follows.

3.4 The five-number summary

Excel commands:

1. Choose **Tools ▶ Data Analysis...**
2. Select **Descriptive Statistics** from the list of **Analysis Tools**
3. Type `TIMES` (or `Data!R2:R21`) in the **Input** range: text box
4. Select the **Output Range** option button and type `E1` for the beginning output cell
5. Select the **Summary Statistics** check box (and deselect the other choices if necessary)
6. Click `OK`

We have reserved the range **E1:F15** for the output. With possible column adjustments for width, your output should resemble that of Figure 3.1.

FIGURE 3.1 Excel's Descriptive Statistics Output

Column1	
Mean	30.25
Standard Error	2.82831082
Median	30.5
Mode	30
Standard Deviation	12.6485905
Sample Variance	159.986842
Kurtosis	2.67701605
Skewness	0.73532243
Range	61
Minimum	5
Maximum	66
Sum	605
Count	20

The first entry is the sample mean of 30.25. Next comes **Standard Error** of 2.82. You will find this discussed at the appropriate time in Chapter 7 of *IS/5e* and *ES/4e*. It is an important statistic related to the standard deviation and is used throughout the implementation of inferential statistics.

Next in order come the Median, Mode, and Standard Deviation. Further down the list you will find the Range, the Minimum and the Maximum for this data set. Finally, the Sum is printed out and the last entry of the output gives the number of data values, which in this case is 20. However, no quartiles are given and, as we have already noted, would differ from the ones used in *IS/5e* and *ES/4e* anyway.

If you will click on the **CH3** worksheet you will see that we have reserved the range **B1:C12** for our version of data summaries. We call this panel **Descriptive Measures**. Here we output the quartiles computed according to Definition 3.7 of *IS/5e* and *ES/4e* as well as other summaries of note. To implement this panel, you have only to type the range or the name of a range of data (even if it is from another worksheet) and the entries will be changed for you immediately. This has an advantage over the **Descriptive Statistics** tool in that the latter must be re-run for every new data set; it is not dynamic. The **Descriptive Measures** panel does operate dynamically; you merely change the name of the data in the input cell to get a new output. We illustrate with the next example.

EXAMPLE 3.10   *Illustrates the use of Descriptive Measures*

Use the Descriptive Measures range in the **CH3** worksheet to obtain several descriptive measures of the viewing-time data given in Table 3.4 on page 37.

SOLUTION   The data are still stored for you in a range named TIMES in the **DATA** worksheet. Click on the **CH3** worksheet tab and either type TIMES or DATA!R2:R21 in the input cell for **DATA FILE:** and the output will be generated immediately. It should resemble the display in Figure 3.2.

FIGURE 3.2
Descriptive
Measures Panel

DATA FILE:	DESCRIPTIVE
TIMES	MEASURES
n =	20
Mean =	30.250
Stdev =	12.649
SE Mean =	2.828
Min =	5.000
Q_1 =	22
Median =	30.500
Q_3 =	37.25
Max =	66.000
IQR =	15.250

You will observe that, apart from rounding, the values in this output agree with those previously obtained. In addition, the quartiles and the interquartile range agree with those published in *IS/5e* and *ES/4e* on page 162. The five-number summary discussed on page 163 of *IS/5e* and *ES/4e* may be found in Rows 7 through 11 ordered from low to high, from Min to Max. ∎

Excel has no option for constructing a boxplot directly. However, with the five-number output of Descriptive Measures, it is a simple matter to construct one using the guidelines outlined in Procedure 3.1 on page 164 of *IS/5e* and *ES/4e*.

PROBLEM 3.9   Use the Descriptive Measures range in CH3 to obtain several descriptive measures, including the five-number summary, for the sample of energy expenditures of Problem 3.8 on page 38 stored in a column named ENERGY.   ◊

3.5 Descriptive measures for populations

We will now see how Excel can be used to determine the mean and standard deviation of a finite population. Those two parameters are the most important descriptive measures of a population.

THE POPULATION MEAN

[E-167 : I-167]

The *population mean*, μ, of a finite population of size N is defined by

$$\mu = \frac{\Sigma x}{N}.$$

Thus we see that the population mean is just the mean of a set of population data. Consequently, to determine a population mean using Excel, we simply employ the AVERAGE function, as indicated in the following example. However, in order to distinguish the population mean from a sample mean, we denote the former using the Greek letter μ and the sample mean, if you will recall, is denoted $\bar{x}$.

EXAMPLE 3.11 *Illustrates finding the population mean*

Table 3.15 on page 176 of *IS/5e* and *ES/4e* presents some data for the 53 active players on the 1997–1998 Dallas Cowboys football team. The weights of these players are summarized here in Table 3.5. Determine the population mean weight of these 11 players with the aid of Excel.

TABLE 3.5
Dallas Cowboys team weights

167	167	219	195	208	202	209	172	198
190	195	219	190	197	225	183	213	189
257	242	294	254	215	230	251	237	248
236	320	308	304	331	291	314	326	278
300	328	170	255	220	190	236	205	207
280	308	263	291	318	267	263	284	

SOLUTION We have stored the weight data into a column named WTS in the **DATA** worksheet. Simply select any empty cell in the Descriptive worksheet and type the function =AVERAGE(WTS) and you will find the value 242.62 displayed. Hence the population mean weight of the players is $\mu = 242.62$ lb. Nothing new here and only the notation μ tells us that the computation represents that of a population mean and not a sample mean. ■

PROBLEM 3.10 The corresponding heights, in inches, for the players in Example 3.11 are given in the table on the following page.

70	70	76	74	75	68	69	69	71
71	71	73	70	71	73	70	73	72
71	74	74	76	70	74	74	74	74
75	75	75	75	79	77	77	75	75
76	78	72	74	77	71	76	73	74
79	79	78	78	78	78	77	77	

Apply Excel to obtain the population mean height of these 53 players. You will find the data stored in a column named HTS for the name of the column containing the data. ◇

THE POPULATION STANDARD DEVIATION   E-175:I-175

Recall that the *population standard deviation*, σ, of a finite population of size N is defined by

$$\sigma = \sqrt{\frac{\Sigma(x-\mu)^2}{N}}.$$

The essential difference in *computation* between the population standard deviation and the sample standard deviation is we divide by *one less* than the number of pieces of data in the formula for the sample standard deviation.

Excel has a special function for computing the population standard deviation instead of the sample standard deviation. The function is called STDEVP and produces σ as the output.

EXAMPLE 3.12 *Illustrates the STDEVP function*

Consider again the weights of the starting offense of the 1997–1998 Dallas Cowboys football team. Those weights are given in Table 3.5 on page 41. Apply Excel to determine the population standard deviation of the weights.

SOLUTION We have already stored the weight data in WTS. Select any empty cell in the Descriptive workbook and type =STDEVP(WTS). You should see the value 48.308 in the cell. Consequently, we see that the population standard deviation of the weights is $\sigma = 48.308$ lb. ∎

PROBLEM 3.11 Refer to Problem 3.10 on the previous page. Use Excel to determine the population standard deviation of the heights of the 53 players. ◇

LESSON 4

Probability Concepts

GENERAL OBJECTIVE In Lessons 2 and 3, we learned how Excel can be used to accomplish the tasks of descriptive statistics. Now we will see how Excel can be applied to perform some of the probability computations discussed in Chapter 4 of *IS/5e* and Chapter 5 of *ES/4e* .

LESSON OUTLINE
4.1 Probability basics
4.2 Contingency tables; joint and marginal probabilities
4.3 Conditional probability

4.1 Probability basics

E-251 : I-201

When the possible outcomes of an experiment are *equally likely* to occur, probabilities are relative frequencies or percentages. Specifically, suppose there are N equally likely outcomes for an experiment. Then the probability an event occurs equals the number of ways, f, that the event can occur, divided by the total number, N, of possible outcomes. That is, the probability equals

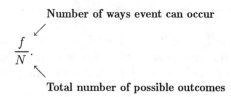

We can use Excel to perform the calculations required in applying the f/N rule, as the following example shows.

EXAMPLE 4.1 *Illustrates how Excel can be used to apply the f/N rule*

The U.S. Bureau of the Census compiles data on family income and publishes its findings in *Current Population Reports*. Table 4.1 gives a frequency distribution of annual income for U.S. families in 1995.

TABLE 4.1 Frequency distribution of annual incomes for U.S. families

Income	Frequency
Under $10,000	5,216
$10,000–$14,999	4,507
$15,000–$24,999	10,040
$25,000–$34,999	9,828
$35,000–$49,999	12,841
$50,000–$74,999	14,204
$75,000 & over	12,961
	69,597

Use Excel to find the probability that a randomly selected family makes
a) between $50,000 and $74,999 inclusive.
b) between $25,000 and $74,999 inclusive.

SOLUTION Open the **Probability** workbook accompanying this manual. The data given in Table 4.1 have been stored in a columns **A** and **B** of the **DATA** worksheet of the **Probability** workbook, with a row for **Totals** added. The total number of families in the city, N, equals the sum of the frequencies in the second column of Table 4.1. Since the frequencies are stored in **Frequency**, we found N with Excel by selecting cell **B9** and clicking the AutoSum button, Σ, in the toolbar. Then we stored each computation of probability by the f/N rule in the column named **Prob** using the simple Excel commands that follow. Recall that when you select a cell and begin typing a formula, the result appears in the formula bar as you type. You must always start with the = sign.

Excel commands:

1. Select cell **C2**, type `=Frequency/SUM(Frequency)` and press Enter
2. Drag the fill handle of cell **C2** down through cell **C8**
3. Select cell **C9** and click the AutoSum button, Σ

The results are shown in Figure 4.1.

FIGURE 4.1 Probability distribution of Y using Excel

	A	B	C
1	Income	Frequency	Prob
2	Under $10,000	5216	0.075
3	$10,000-$14,999	4507	0.065
4	$15,000-$24,999	10040	0.144
5	$25,000-$34,999	9828	0.141
6	$35,000-$49,999	12841	0.185
7	$50,000-$74,999	14204	0.204
8	$75,000 & over	12961	0.186
9	Totals	69597	1.000

a) The probability that a randomly selected family makes between $50,000 and $74,999 may be found in cell **C7** and is listed as 0.204, or about 20%.

b) For this part, we want to find the probability that a randomly selected family makes between $25,000 and $74,999. That probability is the sum of the fifth, sixth, and seventh rows of **Prob**. Select any empty cell in the workbook, type `=C5+C6+C7` and the result is displayed instantly as 0.530. Thus, about 53% of U.S. families made between $25,000 and $74,999 in 1995. ■

PROBLEM 4.1 Apply Excel to the data in Table 4.1 on page 44 to find the probability that a randomly selected family makes
a) less than $15,000.
b) between $10,000 and $34,999.
c) either less than $10,000 or at least $74,999. ◊

4.2 Contingency tables; joint probability distributions

Contingency (or two-way) tables, are used to display frequency distributions for bivariate data for a population. From such a table we can construct various probability distributions and Excel can assist us in that matter. The data displayed in Table 4.2 on the following page are adapted from the *Arizona State University Statistical Summary* (see Table 4.6 on page 229 of *IS/5e*).

TABLE 4.2 Contingency table for age vs. rank of faculty members

		RANK				
		Full Professor R_1	Associate Professor R_2	Assistant Professor R_3	Instructor R_4	Total
A G E	Under 30 A_1	2	3	57	6	68
	30–39 A_2	52	170	163	17	402
	40–49 A_3	156	125	61	6	348
	50–59 A_4	145	68	36	4	253
	60 & over A_5	75	15	3	0	93
	Total	430	381	320	33	1164

The table provides us with a frequency distribution obtained by cross classifying the faculty according to the two characteristics age and rank. In this section we will learn how Excel can be applied to store contingency tables in the computer and to manipulate such tables to obtain joint probability distributions.

CONTINGENCY TABLES IN EXCEL

I-228

We will use Table 4.2 for the next example to illustrate how to store and display a contingency table using Excel.

EXAMPLE 4.2 *Excel contingency table*

Use Excel to store the contingency table, Table 4.2, into the computer. After the table is stored, have Excel display the contingency table.

SOLUTION First, we name three empty, neighboring columns AGE, RANK, FREQ in an Excel worksheet (we chose D, E and F in the **DATA** worksheet of the **Probability** workbook). There are 20 cells (ignoring the Total row and column) containing frequencies for this table. Each cell will occupy a unique AGE, RANK, and FREQ combination in the Excel workspace. If the data were the original raw data, there would be one cell to enter for each of the 1164 cases represented by the table. For example the two Full professors under 30 would each be listed in the original survey as two separate cases (by name probably). Here, that much summary has already been accomplished for you in the table layout. It makes no difference to Excel provided the data are grouped in triplets with the frequency being recorded for a given pair as follows.

4.2 Contingency tables; joint probability distributions

Starting in cell **D2**, we entered the triplet (Under 30, Full,2) in **D2, E2, F2**, respectively. The order of entry of triplets is immaterial but you may find it easier if you proceed row by row or column by column. Thus we entered (Under 30, Assoc, 3) in the third row, but you could enter (30–39, Full, 52) just as well. What is important is that in the end there are 20 rows (**F2** through **F21**) filled with frequencies corresponding to each cell in the table. The data set (including the column names) now occupies the range **D1:F21** and we have named this entire range **FACULTY**.

To construct and display the contingency table we implemented the following steps. First, open the **CHARTS** worksheet (or the **PRACTICE** worksheet if you would prefer doing the steps yourself).

Excel commands:

1 Select **D**ata ➤ **P**ivot Table Report...
2 Select the **M**icrosoft Excel list or data base option and click [Next >]
3 Type **FACULTY** in the **R**ange: text box (or use the button to select the range **D1:F21**) and click [Next >]
4 Drag the [AGE] button into the **R**OW box, drag the [RANK] button into the **C**OLUMN box, drag the [FREQ] button into the **D**ATA box, and click [Next >]
5 Select the **E**xisting worksheet option and use the button to select any cell as the corner of a block of empty cells (we chose **B2**)
6 Select [Finish]

This will produce a table resembling Table 4.6 on page 229 of *IS/5e*. But the order of columns and rows is different since Excel alphabetizes the names. We edited the table to produce the one in Figure 4.2 using the commands following the table.

FIGURE 4.2
Excel contingency table

Sum of FREQ	RANK				
AGE	Full	Assoc	Asst	Instr	Grand Total
Under 30	2	3	57	6	68
30--39	52	170	163	17	402
40--49	156	125	61	6	348
50--59	145	68	36	4	253
60&Over	75	15	3	0	93
Grand Total	430	381	320	33	1164

Excel commands:

1 Click on the name Assoc in cell **C3**
2 Type **Full** and press [Enter ↵]
3 Click on the name 30–39 in cell **B4**
4 Type **Under 30** and press [Enter ↵]

Notice that the colum and/or row shifted to the desired location instantly. We also added some embellishments like borders and shading, but that is not necessary.

Compare this with Table 4.6 in *IS/5e* and you will see that the entries are identical. Totals are identified by the name **Grand Total** in the Excel output. Thus, there are a total of 402 faculty members that are age 30–39. Of these, 170 have the rank of Associate Professor. ∎

PROBLEM 4.2 The following contingency table was obtained by cross-classifying the members of the U.S. House of Representatives (98th Congress) according to years of seniority and political party. [Source: U.S. Congress, Joint Committee on Printing.]

	PARTY		
YEARS	Democrat P_1	Republican P_2	Total
Under 2 S_1	59	25	84
2–9 S_2	134	93	227
10–19 S_3	44	41	85
20–29 S_4	20	8	28
30 & over S_5	10	0	10
Total	267	167	434

Employ Excel to store and display the contingency table. The data are stored in the **DATA** worksheet of the Probability workbook under the name **HOUSE**. ◊

JOINT PROBABILITY DISTRIBUTIONS; MARGINAL DISTRIBUTIONS

Once we have a contingency table stored, it is easy to obtain the joint probability distribution corresponding to the contingency table. Let us return to the situation of Example 4.2 on page 46.

EXAMPLE 4.3 *Illustrates how to obtain joint and marginal probability distributions using Excel*

Recall that Table 4.2 on page 46 provides a contingency table that cross-classifies the faculty at Arizona State University by age and rank. Suppose a faculty member

4.2 Contingency tables; joint probability distributions

is selected at random. Apply Excel to obtain the joint probability distribution for the age versus rank data.

SOLUTION As we see from Table 4.2, there is a total of $N = 1164$ faculty members. Thus, the joint probability distribution for the age versus rank data is obtained by dividing each frequency in Table 4.2 by 1164 according to the f/N rule. Excel will do this easily now that we have the table stored. We first made a copy of the contingency table starting with cell **B11** in the **CHARTS** worksheet using the copy and paste buttons in the toolbar. Then we executed the following steps.

Excel commands:

1. Right click the **Sum of FREQ** cell in the contingency table and select **Wizard...** in the drop down menu
2. Double click the **Sum of FREQ** button in the **DATA** box and type `Joint Probability` in the **Name:** text box
3. Click [Options>>], click ▼ in the **Show data as:** input box, and select **% of total**
4. Click [Number...], select **Number** from the **Category:** drop down list and select **3** from the **Decimal places:** text box
5. Click [OK] twice and then select [Finish]

Your table should resemble the one shown in Figure 4.3.

FIGURE 4.3
Excel joint probability table

Joint Probability	RANK				
AGE	Full	Assoc	Asst	Instr	Grand Total
Under 30	0.002	0.003	0.049	0.005	0.058
30--39	0.045	0.146	0.140	0.015	0.345
40--49	0.134	0.107	0.052	0.005	0.299
50--59	0.125	0.058	0.031	0.003	0.217
60&Over	0.064	0.013	0.003	0.000	0.080
Grand Total	0.369	0.327	0.275	0.028	1.000

The entries in this table are given as probabilities, found by the f/N rule. The probability of drawing a Full professor at random from this population is 0.369 or 36.9% as seen under RANK in the Grand Total row. As noted on page 231 of *IS/5e*, this probability is one item in the marginal distribution corresponding to this row.

Similarly, the probability of drawing a professor at random who is age 50–59 from this population is 0.217 or 21.7% as seen in row **50–59** under the Grand Total column. We may also say that 5.8% of the faculty are Associate professors AND are age 50–59. You should compare the joint probability distribution in Figure 4.2 to the one given in Table 4.7 on page 231 of *IS/5e*. They should be identical except for rounding differences that might come about from the way you have chosen to format the cells in the output. ■

PROBLEM 4.3 Refer to the contingency table in Problem 4.2 on page 48. Suppose a representative is selected at random. Use Excel to find the joint probability distribution for the years of seniority versus political party data. What is the marginal probability of obtaining a congressman age 20–29? ◊

4.3 Conditional probability

[I-240]

On page 238 of *IS/5e*, you learned how to find conditional probabilities from a contingency table. In Excel's **Pivot Table** it is a simple matter to store all such conditional probabilities simultaneously once you have the contingency table stored. Let us illustrate with the next example.

EXAMPLE 4.4 *Illustrates how to obtain conditional probability distributions using Excel*

Determine the conditional probabilities corresponding to each Age classification with **RANK** the conditioning variable using Excel.

SOLUTION We made yet another copy of the contingency table (Figure 4.2)w in the **CHARTS** worksheet starting in cell **I2**, then executed the following steps.

Excel commands:

1. Right click the **Sum of FREQ** cell in the frequency table and select **Wizard...** in the drop down menu
2. Double click the **Sum of FREQ** field in the **DATA** box and type `Conditioned on RANK` in the **Name:** text box
3. Click [Options>>], click [▼] in the **Show data as:** input box, and select **% of column**
4. Click [Number...], select **Number** from the **Category:** drop down list, and select **3** from the **Decimal places:** text box
5. Click [OK] twice and then select [Finish]

Your table should resemble the one given in Figure 4.4.

FIGURE 4.4
Excel conditional (on RANK) probability table

Conditioned on RANK	RANK				
AGE	Full	Assoc	Asst	Instr	Grand Total
Under 30	0.005	0.008	0.178	0.182	0.058
30–39	0.121	0.446	0.509	0.515	0.345
40–49	0.363	0.328	0.191	0.182	0.299
50–59	0.337	0.178	0.113	0.121	0.217
60&Over	0.174	0.039	0.009	0.000	0.080
Grand Total	1.000	1.000	1.000	1.000	1.000

4.3 Conditional probability   51

The entries in this table are conditional probabilities for each given rank. Thus, we can read the conditional probability that a faculty member is in the age 50–59 bracket given that he or she is an Assistant professor directly from the table as 0.113. This is consistent with the computations given on page 238 of *IS/5e*. Here all such conditional probabilities may be read from the table directly.

In addition, the marginal probabilities are maintained in the `Grand Total` column, so that the probability of a faculty member age 50–59 without regard to his or her rank is 0.217, consistent with Figure 4.3 for joint probabilities. ∎

Similarly of course all conditional probabilities conditioned on the AGE variable may be obtained as illustrated in the next example.

EXAMPLE 4.5 *Illustrates how to obtain conditional probability distributions using Excel*

Determine the conditional probabilities for each Rank classification with AGE the conditioning variable using Excel.

SOLUTION Once more we made a copy of the contingency table, this time starting in cell **I11**. We produced Figure 4.5 executing the steps below the figure.

FIGURE 4.5
Excel conditional (on AGE) probability table

Conditioned on AGE	RANK				
AGE	Full	Assoc	Asst	Instr	Grand Total
Under 30	0.029	0.044	0.838	0.088	1.000
30--39	0.129	0.423	0.405	0.042	1.000
40--49	0.448	0.359	0.175	0.017	1.000
50--59	0.573	0.269	0.142	0.016	1.000
60&Over	0.806	0.161	0.032	0.000	1.000
Grand Total	0.369	0.327	0.275	0.028	1.000

Excel commands:

1. Right click the **Sum of FREQ** cell in the frequency table and select **Wizard...** in the drop down menu
2. Double click the **Sum of FREQ** field in the **DATA** box and type `Conditioned on RANK` in the **Name:** text box
3. Click `Options>>`, click ▼ in the **Show data as:** input box, and select **% of row**
4. Click `Number...`, select **Number** from the **Category:** drop down list, and select 3 from the **Decimal places:** text box
5. Click `OK` twice and then select `Finish`

The entries in this table are conditional probabilities. Thus, we can read the conditional probability that a faculty member is an Instructor given that he or she is in the age 30–39 bracket can be read directly from the table as 0.042. All such conditional probabilities may be read from the output directly.

Also, the marginal probabilities are maintained in the `Grand Total` row, so that the probability of a faculty member being an Instructor regardless of his or her age is 0.028, consistent with Figure 4.2 for joint probabilities. ■

PROBLEM 4.4   Refer to the contingency table in Problem 4.2 on page 48. Use Excel to determine the conditional probability of a Republican among congressman who have been in office 30 or more years. Find the conditional probability that a congressman has been in office under 2 years given that he or she is a Democrat. Suppose a representative is selected at random.   ◊

LESSON 5

Discrete Random Variables

GENERAL OBJECTIVE Excel is also useful for dealing with random variables and probability distributions. Recall that a **random variable** is a numerical quantity whose value depends on chance. In this lesson we will learn how Excel can be used to perform many of the computations required for *discrete* random variables.

LESSON OUTLINE
5.1 Discrete random variables and probability distributions
5.2 The mean and standard deviation of a discrete random variable
5.3 The binomial distribution
5.4 The Poisson distribution

5.1 Discrete random variables and probability distributions

E-283 : I-294

A *discrete random variable* is a random variable whose possible values form a discrete data set. Generally, a discrete random variable involves a *count* of something. Examples of discrete random variables are the number of cars owned by a randomly selected family, the number of people waiting for a haircut in a barber shop, and the number of households in a sample that own a color television set.

The *probability distribution* of a discrete random variable consists of a listing of (or formula for) the probabilities associated with the various values of the random variable. Often, the probability distribution of a discrete random variable is obtained from a frequency distribution. For such cases, Excel can be of help in performing the necessary computations.

EXAMPLE 5.1 *Illustrates how Excel can be used to obtain probability distributions*

Table 5.1 gives a frequency distribution for the enrollment by grade level in public elementary schools in the U.S. [Source: U.S. National Center for Education Statistics]. (0 = kindergarten, 1 = first grade, and so on). Frequencies are in thousands of students.

TABLE 5.1
Frequency distribution for enrollment by grade in U.S. public elementary schools

Grade level y	Frequency f
0	4,043
1	3,593
2	3,440
3	3,439
4	3,426
5	3,372
6	3,381
7	3,404
8	3,302

Suppose a student in elementary school is to be selected at random. Let Y denote the grade of the student obtained. Employ Excel to determine the probability distribution of the random variable Y.

SOLUTION To obtain the probability distribution of Y we need to apply the f/N rule to each of the frequencies in the second column of Table 5.1. We entered the data of Table 5.1 into the **DATA** worksheet of the **Probability** workbook into two columns labeled y and f (columns **L,M**). Then we executed the steps on the following page, remaining in the **DATA** worksheet for the sake of juxtaposition. (Of course you may repeat these steps in the **PRACTICE** worksheet if you wish.)

Excel commands:

1 Click on cell **N2**
2 Type `=f/sum(f)` and press Enter↵
3 Drag the fill handle of **N2** down through cell **N8**

The results are displayed immediately and should resemble those in Figure 5.1

FIGURE 5.1
Excel display of the probability distribution of Y

y	f	P(Y=y)
0	4043	0.128758
1	3593	0.114427
2	3440	0.109554
3	3439	0.109522
4	3426	0.109108
5	3372	0.107389
6	3381	0.107675
7	3404	0.108408
8	3302	0.105159

You should compare the probability distribution in Figure 5.1 to the one you will find in Table 5.4 on page 294 of *IS/5e* (Table 5.9 on page 283 of *ES/4e*). ■

In the future, you may use these three columns for similar problems, writing over the current results and deleting any that might occur in a given column below the entry of new data.

PROBLEM 5.1 According to the U.S. Bureau of the Census, a frequency distribution for the number of persons per household in the U.S. is as shown in the table below. Frequencies are in millions.

Number of persons y	Number of households f
1	19.4
2	26.5
3	14.6
4	12.9
5	6.1
6	2.5
7	1.6

Suppose a U.S. household is to be selected at random. Let Y denote the number of persons in the household chosen. Employ Excel to obtain the probability distribution of the random variable Y. Use y and f for storing the data. ◊

INTERPRETATION OF PROBABILITY DISTRIBUTIONS

E-285 : I-297

The relative frequency interpretation of probability implies that, in a large number of independent repetitions of the experiment, the relative frequency of occurrence of an event will be close to the probability of that event. We can use Excel to mimic independent repetitions of an experiment involving the observation of a random variable using a special **Random Number Generator** program in the **Data Analysis** tool. This process is referred to as *simulation* and is a very powerful tool using random number generators.

Among the various choices of distributions to generate from, one is called DISCRETE. This choice allows us to simulate observations from any discrete distribution stored in a pair of neighboring columns in an Excel worksheet. We will illustrate this with the next example.

EXAMPLE 5.2 *Illustrates the use of Excel's Random Number Generator*

In Table 5.2 that follows, we display the probability distribution for X, the number of heads obtained in three tosses of a balanced dime, duplicating Table 5.6 on page 295 of *IS/5e* (Table 5.11 on page 284 of *ES/4e*).

TABLE 5.2 Probability distribution of the random variable X, the number of heads obtained in three tosses of a balanced dime

No. of heads x	Probability $P(X = x)$
0	0.125
1	0.375
2	0.375
3	0.125

Use Excel to simulate 1000 observations of the number of heads obtained in three tosses of a balanced dime and compare the relative frequencies with the probabilities.

SOLUTION First, store this probability distribution in columns labeled x and P(X=x) in the **DATA** worksheet of the Probability workbook in the range **Q2:R5**, deleting any entries below Row 5 in those columns if necessary. Then carry out the simulation executing the following steps.

Excel commands:

1 Choose **Tools ▶ Data Analysis**...
2 Select **Random Number Generation** from the **Analysis Tools** menu box and click [OK]

3 Select the **Number of Variables:** text box and type 1
4 Select the **Number of Random Numbers:** text box and type 1000
5 Select **Discrete** from the **Distribution:** text box
6 Use the [📷] button to select the range Q2:R5 for the **Value and Probability Input Range:** text box
7 Select the **Output Range:** option, select its text box and use the [📷] button to select the output range **AG2**
8 Click [OK]

This much collects 1000 independent observations of this random variable and stores them in the column SAMPL (**AG**) in the **DATA** worksheet. We need to process these results to discover how many times each observation occurred and the relative frequency of occurrence. We can do this with Excel's COUNTIF function. This function takes two arguments, a range of cells (or name) it is to act upon and a *criterion* to satisfy. In the first cell we direct Excel to count the observation only if it is 0 in SAMPL. Then we will have to adjust for the other values of the random variable. Finally, we will convert counts to relative frequencies as in the following steps.

Excel commands:

1 Click on **S2** and type =Countif(SAMPL,="0")
2 Click on **T2** and type =COUNT/SUM(COUNT)
3 Select cells **S2** and **T2** and drag the fill handle of **T2** down through cell **T5**
4 Manually replace ="0" with ="1", ="2", and ="3" in the respective cells **T3**, **T4**, and **T5**

Our results may be viewed in Figure 5.2.

FIGURE 5.2
Excel output for 1,000 simulated trials

x	P(X=x)	Count	Rel Freq
0	0.029	130	0.130
1	0.049	366	0.366
2	0.078	363	0.363
3	0.155	141	0.141

Notice the somewhat close agreement between the probabilities in Table 5.2 and the relative frequencies found here. Of course even with 1,000 replications we expect to see differences. And, indeed, if we were to perform this simulation a second time, we would not expect to see the same results. ■

PROBLEM 5.2 Using the functions illustrated in Example 5.2, generate your own sample of 1000 independent observations of the distribution in Table 5.2. Summarize the results and compare the relative frequencies to the probabilities. ◊

5.2 Mean and standard deviation of a discrete random variable

E-295 : I-307

In this section we will learn how Excel can be used to determine the mean and standard deviation of a discrete random variable. We first consider the mean of a discrete random variable.

The *mean*, μ, of a discrete random variable X is defined by

$$\mu = \Sigma x P(X = x)$$

and the *standard deviation*, σ, of X is defined by

$$\sigma = \sqrt{\Sigma(x)^2 P(X = x) - \mu^2}.$$

It is an extremely simple matter in Excel to compute both the mean and standard deviation of a discrete random variable.

In the **DATA** worksheet in the Excel Probability workbook, the columns labeled x and P(X=x) (**Q** and **R**) have been reserved for storing the probability distribution of any random variable X. In those columns, rows 2 through 51 have been reserved for data entry and are named x and Px. This allows you to enter up to 50 pairs of values for any distribution you wish and call them within Excel by name. Let us illustrate.

EXAMPLE 5.3 *Illustrates how to use Excel to compute the mean and standard deviation of a discrete random variable*

Prescott National Bank has six tellers available to serve customers. The number of tellers busy with customers at, say, 1:00 P.M. varies from day to day and depends on chance; so it is random variable, which we will call X. Past records indicate that the probability distribution of X is as shown in the Table 5.3. Employ Excel to determine the mean, μ, of the random variable X.

TABLE 5.3 Probability distribution of the random variable X, the number of Tellers busy with customers

x	P(X = x)
0	0.029
1	0.049
2	0.078
3	0.155
4	0.212
5	0.262
6	0.215

SOLUTION Enter the probability distribution shown in Table 5.3 into the columns labeled x and P(X=x) in the **DATA** worksheet of the Probability workbook as in the last example.[†]

[†] Be sure to clear any entries below your last one if there are any.

In that worksheet, cells **O2** and **P2** (labeled μ and σ) have been programmed as follows for recording the mean and standard deviation of a discrete random variable of any distribution entered in columns x and P(X=x).

Excel commands:

1. Select cell **O2**, type =SUMPRODUCT(x,Px) and press [Enter↵]
2. Select cell **P2**, type =SQRT(SUMPRODUCT(x^2,Px)-mu^2) and press [Enter↵]

The mean and standard deviation of X are computed as soon as a probability distribution is entered in columns **Q** and **R**. Here, you will find the mean and standard deviation recorded as $\mu = 4.118$ and $\sigma = 1.575$. ■

PROBLEM 5.3 Refer to Problem 5.1 on page 55. In that problem you used Excel to determine the probability distribution of the random variable X—the number of persons in a randomly selected household. Now apply Excel to the probability distribution you obtained to find the mean, μ, and standard deviation, σ, of X. ◊

5.3 The binomial distribution [E-308 : I-321]

Suppose that n Bernoulli trials are performed, with the probability of success on any given trial being p. Let X denote the total number of successes in the n trials. Then probabilities for the random variable X are given by the **binomial probability formula**

$$P(X = x) = \binom{n}{x} p^x (1-p)^{n-x},$$

where $x = 0, 1, \ldots, n$. We refer to n and p as *parameters*.

The computations required for obtaining binomial probabilities using the binomial probability formula can be tedious and time-consuming. Fortunately, Excel has a general function, **BINMODIST** for computing various binomial probabilities. The general format is BINOMDIST(k,n,p,logical), where k is the value of X sought after, n is the number of Bernoulli trials, p is the success probability, and **logical** is to be replaced by **false** to find individual probabilities and **true** to compute cumulative probabilities. We illustrate with the next example.

EXAMPLE 5.4 *Illustrates the BINOMDIST function in Excel*

According to tables provided by the U.S. National Center for Health Statistics in *Vital Statistics of the United States*, there is about an 80% chance that a person age 20 will be alive at age 65. Suppose three people age 20 are selected at random. Compute the following probabilities using the Excel function **BINOMDIST**. For each case, the probability refers to the number alive at age 65.

Lesson 5 Discrete Random Variables

a) exactly two.
b) at most one.
c) at least one.
d) Determine the probability distribution for the number alive at age 65.
e) Construct a probability histogram for the number alive at age 65.

SOLUTION In this situation, a success is a person currently age 20 who is alive at age 65. The probability of a success is the probability that a person currently age 20 will be alive at age 65. This is 80% so that $p = 0.8$. The number of trials is the number of people in the study so $n = 3$. Consequently, the appropriate binomial distribution is the one with parameters $n = 3$ and $p = 0.8$.

a) To obtain the probability that exactly two of the three people will be alive at age 65, two successes, we proceed as follows.
Excel commands:

1 Select any empty cell in any worksheet and type
=BINOMDIST(2,3,0.8,FALSE)

The resulting output will appear instantly in the cell as having the value 0.384. Therefore the probability is 0.3840 that exactly two of the three people will be alive at age 65.

b) Here we want the probability that at most one person will be alive at the age of 65, that is, less than or equal to one. That is, we want the cumulative probability for the value 1. Proceed as follows.
Excel commands:

1 Select any empty cell in any worksheet and type
=BINOMDIST(1,3,0.8,TRUE)

You see that the probability of at most one person being alive at age 65 is 0.104.

c) The event at least one person will be alive at the age of 65 is the complement of the event no one will be alive at the age of 65. Hence the required probability is $1 - P(X = 0)$. Now $P(X = 0)$ may be computed exactly as in part (a) with 0 substituted for 2. The result is 0.008. Consequently, the probability of at least one sale is $1 - 0.008 = 0.992$.

d) Here we are to determine the probability distribution for the number of sales. In other words, we want to obtain the binomial distribution with parameters $n = 3$ and $p = 0.8$. First, open the **DATA** worksheet and store the integers 0–3 in the column labeled x. Clear any rows below this entry in columns **Q** or **R**. With this much accomplished we are ready to store the binomial probabilities. Proceed as follows.

Excel commands:

1 Click on **R2** and type =BINOMDIST(x,3,0.8,FALSE)
2 Select cell **R2** drag its fill handle down through cell **R5**

5.3 The binomial distribution

The results are displayed in Figure 5.3.

FIGURE 5.3
Excel output of the binomial distribution

x	P(X=x)
0	0.008
1	0.096
2	0.384
3	0.512

You should compare this output with Table 5.15 on page 315 of *IS/5e* (Table 5.20 on page 302 of *ES/4e*). As a bonus you should observe that the mean and standard deviation were automatically computed and stored in columns **O** and **P** under μ and σ as well. Indeed, $\mu = 2.4$ and $\sigma = 0.693$.

e) Now that we have the entire binomial probability distribution stored, we may use the Excel Chart Wizard to obtain a probability histogram of the distribution. After some re-sizing and formatting, we were able to produce the histogram displayed in Figure 5.4 using the Excel commands following the figure.

FIGURE 5.4
Excel probability histogram for the Binomial 3, 0.8 distribution

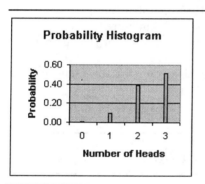

Open the **CHARTS** worksheet of the Probability workbook and select cell **P2**. Then proceed as follows.

Excel commands:

1. Select the range **R2:R5** in the **DATA** worksheet and click the chart wizard button on the toolbar
2. Choose **Column** under **Chart Type:** and click **Next >**
3. Click the **Series** tab and use the button in the **Category (X) axis labels:** text box to select the range **Q2:Q5** in the **DATA** worksheet
4. Click **Next >**, click the **Titles** tab and type Probability Histogram in the **Chart title:** text box
5. Select the **Category (X) axis:** text box and type Number of Heads
6. Select the **Value(Y) axis:** text box and type Probability

7 Click the **Legend** tab, deselect the **Show Legend** check box and click [Next >]
8 Select the **As object in:** radio button, click [▼] in the text box and select **CHARTS**
9 Select [Finish]

Format the resulting chart after sizing it to cover about 10 rows as follows.

Excel commands:

1 Right click on any bar, choose **Format Data Series...**, click the Options tab and set **Gap width** to 500
2 Format the **Value Axis** to a number with 2 decimal places

You should compare this figure with the probability histogram given in Figure 5.15 on page 315 of *IS/5e* (Figure 5.25 on page 303 of *ES/4e*). From any of the displays, the left skewness is readily apparent. ■

PROBLEM 5.4 According to a 1980 telephone study in Phoenix, the probability is about 0.25 that a randomly selected phone call will last longer than the mean duration of 3.8 minutes. What is the probability that, out of *three* randomly selected calls,

a) exactly two last longer than 3.8 minutes?
b) at most two last longer than 3.8 minutes?
c) at least one lasts longer than 3.8 minutes?
d) Determine the probability distribution for the number of calls out of 3 that last longer than 3.8 minutes.
e) Construct a probability histogram for the number of calls, and identify the skewness of this binomial distribution. ◊

5.4 The Poisson distribution [I-336]

The Poisson distribution is regularly used to model the frequency with which a specified event occurs during a particular period of time, as well as a variety of other random phenomena.

The computations required for obtaining Poisson probabilities using the Poisson probability formula is complicated slightly by the need to compute the factorial function $x!$ and the fact that there are, technically, infinitely many possible values of the random variable. Fortunately, Excel has a POISSON function for computing various probabilities when the parameter λ (called **mean** in Excel) is specified.[†] Similar to the binomial distribution, the general format is POISSON(k,λ,logical), where k is the value of X sought after, λ is the mean of X, which must be known, and logical is to be replace by **false** to find individual probabilities and **true** to compute cumulative probabilities. Here is an example of how this program can be used.

[†] As you will see in the next section, λ is the mean, μ, of X. You should know that in many other textbooks, λ is defined to be the reciprocal $1/\mu$ and you must exercise some care in implementing this Excel function in various sets of circumstances.

EXAMPLE 5.5 Illustrates the POISSON function in Excel

Desert Samaritan Hospital keeps detailed records of emergency-room traffic. From these records, we find the number of patients arriving between 6:00 PM and 7:00 PM has a Poisson distribution with parameter $\lambda = 6.9$. Use Excel to compute the following probabilities:

a) exactly four patients arrive.
b) at most two patients arrive.
c) Determine the probability distribution for the number of patients arriving. Only use values for which the probability is positive to three decimal places.
d) Use part (c) to construct a (partial) probability histogram for the random variable X.

SOLUTION Let X denote the number of patients arriving between 6:00 PM and 7:00 PM. Then the random variable X has a Poisson distribution with parameter $\lambda = 6.9$.

a) To obtain the probability of exactly four patients arriving, we proceed as follows.

Excel commands:

1 Select any empty cell in any workspace and type =POISSON(4,6.9,FALSE)

The resulting output will appear instantly in the cell as having the value 0.095. Therefore the probability is 0.095 that exactly four patients will arrive during the time interval specified.

b) Here we want the probability that at most two patients arrive between 6:00 PM and 7:00PM; that is, less than or equal to two arrivals. So, here we want the cumulative probability for the value 2.

Excel commands:

1 Select any empty cell in any workspace and type =POISSON(2,6.9,TRUE)

The resulting output will appear instantly in the cell as having the value. Consequently, the probability of at most two arrivals is 0.0320.

c) Next, we are to determine the probability distribution for the number of arrivals. In other words, we want to obtain the Poisson distribution with parameter $\lambda = 6.9$. Since there are, technically, infinitely many values of X possible, we need to approximate this probability distribution.

Open the **DATA** worksheet and proceed as follows. First, store the integers 0–20 in the column labeled x. Clear any rows below this entry in both columns **Q** and **R**. With this much accomplished we are ready to store the Poisson probabilities. Proceed as follows.

Excel commands:

1 Click on **R2** and type =POISSON(x,6.9,FALSE)
2 Select cell **R2** drag its fill handle down through cell **R22**

Observe that the first value for which the Poisson probability is 0 to three decimal places is at $x = 18$. Thus, we need not consider the distribution beyond this point. As a bonus you will find the mean and standard deviation automatically computed and stored in **O** and **P** under μ and σ. You may confirm Formula 5.4 on page 335 of *IS/5e* by observing that $\mu = 6.900$ and $\sigma = 2.626$.

d) Now we may use the Excel Chart Wizard to obtain a (partial) probability histogram of the distribution proceeding as in the Binomial case. After some resizing and formatting, we were able to produce the histogram displayed in Figure 5.5 using the Excel commands following the figure.

FIGURE 5.5 Excel (partial) probability histogram for the Poisson 6.9 distribution

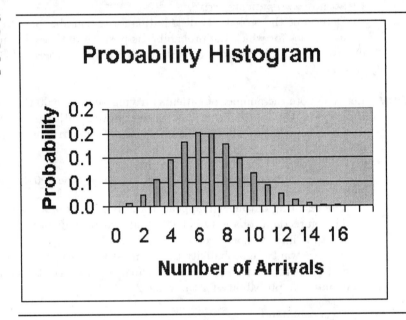

Open the **CHARTS** worksheet of the Probability workbook and select cell **U2**. Then proceed as follows.

Excel commands:

1. Select the range **R2:R20** in the **DATA** worksheet and click the chart wizard button on the toolbar
2. Choose **Column** under **Chart Type:** and click [Next >]
3. Click the **Series** tab and use the button in the **Category (X) axis labels:** text box to select the range **Q2:Q18** in the **DATA** worksheet
4. Click [Next >], click the **Titles** tab and type `Probability Histogram` in the **Chart title:** text box
5. Select the **Category (X) axis:** text box and type `Number of Arrivals`
6. Select the **Value(Y) axis:** text box and type `Probability`
7. Click the **Legend** tab, deselect the **Show Legend** check box and click [Next >]
8. Click ▼ in the **As object in:** drop down box and select **CHARTS**
9. Select [Finish]

As you can see, this particular distribution is skewed slightly to the right. Most of the probability is associated with relatively small values of x with a trailing off to the right. On the other hand, there is a great deal of symmetry in the range of x-values 0–16, where most of the probability is located. Indeed, $P(x > 13) = 0.001$. You may verify this in Excel using the POISSON function as in part (b) in this example. You would discover directly that $P(x \leq 16) = 0.999$ if you type POISSON(16,6.9,TRUE) in any open cell in Excel. ∎

PROBLEM 5.5 Assume that the sales made in a week by a used car salesperson occur like a Poisson random variable with parameter $\lambda = 1$. What is the probability that the salesperson makes,
a) exactly two sales in a week?
b) at most two sales in a week?
c) Determine the probability distribution for the number sales in a week.
d) Use the results to plot the distribution and discuss its shape.   ◇

POISSON APPROXIMATION TO THE BINOMIAL DISTRIBUTION

`I-338`

When n is large and p is small, the Poisson formula with $\lambda = np$ for probabilities can be used to approximate the binomial probabilities having parameters n and p. A general rule of thumb in order to apply the approximation are the two conditions $n \geq 100$ and $np \leq 10$. Let us illustrate with the next example.

EXAMPLE 5.6 *Illustrates the Poisson approximation*

According to the *World Almanac*, the infant mortality rate in Sweden is 4.5 per 1000 live births. Determine the probability that, out of 500 randomly selected live births, there are
a) no infant deaths.
b) at most three infant deaths. Use Excel and the Poisson approximation to compute these probabilities.
c) Compare the results with the corresponding binomial probabilities using Excel's BINOMIAL function.

SOLUTION Clearly the problem as stated is a binomial model with $n = 500$ and $p = 0.0045$. This is a case where the Poisson approximation applies, with $np = 2.25$.
a) To obtain the probability of no infant deaths, we proceed as follows.

Excel commands:

1 Select any empty cell in any workspace and type =POISSON(0,2.25,FALSE)

The result appears immediately in the cell as 0.105. Therefore the probability is 0.105 that there are no infant deaths.

b) Here we need to find the cumulative probability.

Excel commands:

1 Select any empty cell in the workspace and type =POISSON(3,2.25,TRUE)

The result appears immediately in the cell as 0.809. Consequently, the probability of at most three infant deaths is 0.809.

c) Finally, we are to determine these probabilities using the BINOMIAL function. The procedure is straightforward.

Excel commands:

1 Select any empty cell in the workspace and type
   =BINOMDIST(0,500,.0045,FALSE)

The result appears immediately in the cell as 0.105.

Excel commands:

1 Select any empty cell in the workspace and type
   =BINOMDIST(3,500,.0045,TRUE)

The result appears immediately in the cell as 0.810.

Notice that these more exact probabilities are approximated satisfactorily by the Poisson approximate values. However, it was a simple matter to compute the more exact values, making the approximating values redundant. ∎

PROBLEM 5.6 Using Example 5.6 as a model, compute the probabilities that there will be
a) exactly two infant deaths,
b) at most four infant deaths, using both the Poisson approximation formulas and the more exact Binomial formulas. ⋄

LESSON 6

The Normal Distribution

GENERAL OBJECTIVE Of all the probability distributions used in the study of both probability and statistics, the **normal distribution** is the most important. It is often appropriate to use a normal distribution as the distribution of a population or random variable. Furthermore, the normal distribution is applied extensively in inferential statistics. In this lesson we will learn how Excel can be employed to perform a variety of tasks related to the normal distribution.

LESSON OUTLINE
6.1 Normally distributed random variables
6.2 Areas under the standard normal curve
6.3 Working with normally distributed random variables
6.4 Normal probability plots

6.1 Normally distributed random variables [E-337 : I-361]

We may use Excel's Random Number Generator to simulate a normally distributed random variable. We can then plot a histogram of the resulting data and compare the result to a normal curve having the same parameters, namely mean and standard deviation. The following example shows how Excel can be used to accomplish this.

EXAMPLE 6.1 *Illustrates the simulation of a normal distribution*

Gestation periods of humans are normally distributed with a mean of 266 days and a standard deviation of 16 days. Use Excel to simulate 1000 human gestation periods and then obtain a histogram of the results, comparing the histogram to a normal curve having $\mu = 266$ and $\sigma = 16$.

SOLUTION We use the Random Number Generator in Excel to simulate the actual performance of the experiment involving observing the value of X. Open the DATA worksheet of the Probability workbook and proceed as follows.

Excel commands:

1. Choose Tools ➤ Data Analysis...
2. Select Random Number Generation from the Analysis Tools drop down menu and click [OK]
3. Type 1 in the Number of Variables: text box
4. Select the Number of Random Numbers: text box and type 1000
5. Click the ▼ button in the Distribution: text box and select Normal from drop down list
6. Select the Mean = text box and type 266
7. Select the Standard Deviation = text box and type 16
8. Click the Output range option button, select the Output Range: text box and type U2 (or use the ▦ button to select the output range)
9. Click [OK]

This much collects 1000 independent observations of this random variable and stores them in the column SAMPLE in the DATA worksheet. In addition, you will observe (as in the CH2 worksheet of the Descriptive workbook) that the maximum and minimum value of the sample has been recorded in column W, along with the interval width needed to have 5 intervals (21.0 in our case) and the interval width (5.3 in our case) needed to have 20 intervals. We will compromise and use an interval width of 10. We use this information to establish the interval widths, or bins as they are called in Excel, as follows. We will set the first lower cutpoint just enough to include the minimum observed value in cell W2 (210.2 in our case) in the first interval. Then we will want to have the last interval be sure to cover the maximum observed value in cell **W3** (315.3 in our case). Of course your data will differ from these so you might be selecting different limits. Hence the following steps may be viewed as a guideline given the values that we obtained.

Excel commands:

1 Select input cell **V9** and type 210
2 Select input cell **V10** and type 220
3 Select the range **V9:V10** and drag the fill handle for **V10** down to cell **V21** to hold the value 320

Now that we have the choice of bin values, we may proceed to construct the histogram as follows. We produced the histogram shown in Figure 6.1 using the Excel commands below the figure.

FIGURE 6.1
Histogram of 1000 simulated gestation periods

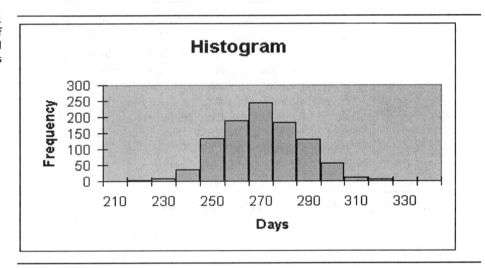

Excel commands:

1 Choose **Tools** ➤ **Data Analysis**...
2 Select **Histogram** from the **Analysis Tools** drop down menu
3 Type SAMPLE in the **Input Range:** text box
4 Use the ▥ button to select **W10:W21** from the **DATA** worksheet for the **Bin Range:** text box
5 Deselect the **Labels** check box
6 Click the **Output Range:** radio button, click on the **CHARTS** tab and type AA2 (or use the ▥ button) to select the output range
7 Click [OK]

We modified the output slightly by formatting the data series to have a gap width of 0 (see Lesson 2, page 25) and we formatted the x-axis to have 2 categories between tick mark labels and no decimal places. We also labeled the x-axis **Days** instead of **Bins** and replaced the word **More** with 330.0.

As expected, the histogram is roughly bell-shaped, the resemblance to a normal distribution being rather pronounced. Naturally, your histogram will not be identical to this one due to the nature of simulation. ∎

PROBLEM 6.1 Using the circumstances of Example 6.1, simulate 1000 gestation periods and use Excel to construct a histogram of the results. ◊

6.2 Areas under the standard normal curve E-343:I-367

Probabilities for normally distributed populations or random variables are equal to areas under a suitable "bell-shaped" curve. Such a curve is called a *normal curve*. There are many, in fact, infinitely many normal curves. In this section we will consider a special normal curve—the *standard normal curve* or *z-curve*.

Areas under the standard normal curve can be determined with the aid of standard normal tables like Table II found in *IS/5e* and *ES/4e*. Alternatively, we can use Excel to find areas under the standard normal curve. The appropriate Excel function is NORMSDIST. This function computes the area under the standard normal curve that lies to the *left* of a specified value of z, that is, a z-score. The general format is NORMSDIST(z) where z is the specified value.

EXAMPLE 6.2 *Illustrates the NORMSDIST function*

Use Excel to find the area under the standard normal curve that lies
a) to the left of $z = 1.23$.
b) to the right of $z = 0.76$.
c) between $z = -0.68$ and $z = 1.82$.

SOLUTION a) As we pointed out earlier, the function NORMSDIST computes the area under the standard normal curve to the left of a z-score.

Excel commands:

1 Click in any empty cell of any worksheet (formatted to 4 decimal places) and type =NORMSDIST(1.23)

The value 0.8907 will appear instantly in the selected cell. Thus, the area under the standard normal curve to the left of $z = 1.23$ is 0.8907.

b) Here we want to find the area under the standard normal curve to the right of $z = 0.76$. Since NORMSDIST computes areas to the *left* of specified z-scores, we proceed as follows.

6.2 Areas under the standard normal curve

Excel commands:

1. Click in any empty cell of any worksheet (formatted to 4 decimal places) and type =1-NORMSDIST(0.76)

The value 0.2236 will appear instantly in the selected cell. Thus, the area under the standard normal curve to the right of $z = 0.76$ is 0.2236.

c) The area under the standard normal curve between $z = -0.68$ and $z = 1.82$ equals the area to the left of $z = 1.82$ *minus* the area to the left of $z = -0.68$. We can use NORMSDIST to obtain these latter two areas as follows.

Excel commands:

1. Click in any empty cell of any worksheet (formatted to 4 decimal places) and type =NORMSDIST(1.82)-NORMSDIST(-.68)

The value 0.7174 will appear instantly in the selected cell. Thus, the area under the standard normal curve between -0.68 and $+1.82$ is 0.7174.

The area under the standard normal curve between any two z-scores can be determined in the same way. ∎

PROBLEM 6.2

Use Excel to find the area under the standard normal curve that lies
a) to the left of $z = -1.28$.
b) between $z = 1.53$ and $z = 2.47$.
c) to the right of $z = 1.64$.
d) either to the left of $z = -1.85$ or to the right of $z = 0.29$. ◊

FINDING THE z-SCORE FOR A SPECIFIED AREA `E-344 : I-368`

Excel also has a function, called NORMSINV, that reverses the activity of NORMSDIST. It corresponds to going into the body of Table II of *IS/5e* or *ES/4e* and finding the value of z—the inverse of finding the area for a given value of z. Since the function NORMSDIST command yields areas all the way to the left of z, the format NORMSINV(probability) requires that you enter such areas as a probability so that percentages must be expressed in decimal form.

EXAMPLE 6.3   *Illustrates how to find the z-score for a specified area using Excel*

Use Excel to find the z-score for which the area under the standard normal curve to the left of that value is 0.04.

SOLUTION Since the area to the left is specified, this is the proper entry for the NORMSINV.

Excel commands:

1 Click in any empty cell of any worksheet (formatted to 2 decimal places) and type =NORMSINV(0.04)

The value -1.75 will appear instantly in the selected cell. This would be denoted $z_{0.96}$ in the standard notation and $z_{0.96} = -z_{0.04}$ from the symmetry property of the normal curve. Note that, for all choices of $\alpha > 0.5$, the value of z_α will be negative. ∎

EXAMPLE 6.4 *Illustrates how to find the z-score for a specified area using Excel*

Use Excel to find the z-value for which the area under the standard normal curve to the right of that value is 0.025.

SOLUTION To find the value using Excel, we must input the area to the *left*, which is 0.975. Proceed as follows.

Excel commands:

1 Click in any empty cell of any worksheet (formatted to 2 decimal places) and type =NORMSINV(0.975)

The value 1.96 will appear instantly in the selected cell. You recognize this as $z_{0.025}$ as defined in Definition 6.3 on page 370 of *IS/5e* (page 346 of *ES/4e*). An input of 0.025 in NORMSINV would produce -1.96 as the output. ∎

In this way you may find the value of z_α for any choice of α. But you must remember to input $1 - \alpha$, not α into the NORMSINV function.

EXAMPLE 6.5 *Illustrates how to find the z-scores for a specified area using Excel*

Use Excel to determine the two z-scores that divide the area under the standard normal curve into a middle 0.95 area and two outside 0.025 areas. See Figure 6.2.

FIGURE 6.2
The standard normal curve

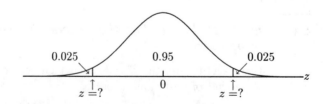

SOLUTION As we see from Figure 6.2, the area of the small region on the right-hand side is 0.025. This means that the z-score on the right is $z_{0.025}$. In Example 6.3, we found that $z_{0.025} = 1.96$. Thus the z-score on the right is 1.96. Since the standard normal curve is symmetric about 0, it follows that the z-score on the left is -1.96. ∎

PROBLEM 6.3 For the standard normal curve
a) find the z-value with area 0.75 to its left.
b) find the z-value with area 0.95 to its right.
c) find the two z-values that divide the area into a middle 0.90.
d) find $z_{0.06}$ and $z_{0.94}$. ◇

6.3 Working with normally distributed random variables

Like the function NORMSDIST, Excel has a function NORMDIST (without the S) that computes areas to the *left* of a specified value. The function has the format NORMDIST(x,μ,σ,true), and returns the area to the left of x for a normal curve having mean μ and standard deviation σ. Similarly, the Excel function with format NORMINV(probability, μ, σ) will return the x-value for which the area to the left of x is probability, under a normal curve having mean μ and standard deviation σ.

AREAS UNDER ANY NORMAL CURVE E-353:I-377

On page 375 of *IS/5e* (page 352 of *ES/4e*) you learned a procedure for finding a percentage or probability for any normally distributed variable. Excel makes it unnecessary to calculate the intermediate z-scores to find such areas, (Step 3 of that procedure) since the function NORMDIST that will allow for any mean μ and standard deviation σ. Here is an example.

EXAMPLE 6.6 *Illustrates finding a probability for a normally distributed variable*

Intelligence quotients (IQ's) measured on the Stanford Revision of the Binet-Simon Intelligence Scale are known to be normally distributed with a mean of 100 and a standard deviation of 16. Obtain the percentage of people having IQs between 115 and 140.

SOLUTION We want to find the area under the normal curve with parameters $\mu = 100$ and $\sigma = 16$ that lies between 115 and 140. Proceed as follows.

Excel commands:

1 Click in any empty cell of any worksheet (formatted to 4 decimal places) and type =NORDIST(140,100,16,true)-NORDIST(115,100,16,true)

74   Lesson 6   The Normal Distribution

The value 0.1680 will appear instantly in the selected cell. Consequently, we see that approximately 16.8% of the population have IQs between 115 and 140. ∎

PROBLEM 6.4 Use Excel to find the area under the normal curve with parameters $\mu = 15$ and standard deviation $\sigma = 2$ that lies
a) to the left of 12.
b) between 14 and 18.
c) to the right of 16.5.   ◊

FINDING THE x-VALUE FOR A SPECIFIED PROBABILITY

E-356 : I-380

For this general problem we need to invoke the function NORMINV, with specification of arbitrary parameters. That means you can find the x-value corresponding to any area to the left of that point, just as you were able to find the z-score in the standard normal curve. In particular you may find percentiles for any normal curve.

EXAMPLE 6.7 *Illustrates finding the x-value for a specified probability*

For the circumstances of Example 6.6, find the 90th percentile for IQs.

SOLUTION We need to find the x-value corresponding to an area of 0.90 to the left of that value. Proceed as follows.

Excel commands:

1 Click in any empty cell of any worksheet (formatted to 2 decimal places) and type =NORMINV(0.90,100,16)

The value 120.5 will appear instantly in the selected cell. Thus, the 90th percentile is about 120.5. In other words, approximately 90% of the population has an IQ less that 120.5. ∎

PROBLEM 6.5 For the normal curve with parameters $\mu = 64.6$ and $\sigma = 2.4$, use Excel to find the x-value with
a) area 0.90 to its left.
b) area 0.90 to its right.   ◊

6.4 Normal probability plots

E-363 : I-387

Normal probability plots are useful tools for indicating whether it is reasonable to assume that the population under discussion can be represented by a normal

6.4 Normal probability plots

distribution. Excel can be used to convert a set of data to normal scores for you. Then we may use the CHART WIZARD to produce a normal probability plot. We illustrate this with the next example.

EXAMPLE 6.8 *Illustrates the construction of Probability Plots*

In Table 6.1, we repeat the data for 12 IRS returns, as discussed in Example 6.14 on page 386 in *IS/5e* (page 362 of *ES/4e*).

TABLE 6.1
Adjusted gross incomes ($1000)

9.7	93.1	33.0	21.2
81.4	51.1	43.5	10.6
12.8	7.8	18.1	12.7

Use Excel to construct a probability plot for the data.

SOLUTION We stored the data in a column labeled AGI in the DATA worksheet of the Probability notebook. First, open the NSCORES worksheet in that workbook. There you will find the framework for computing Nscores from raw data. The first three columns have been reserved for this purpose.

The first column, labeled RANK, contains the integers 1–100 to allow for that large a sample size. You may extend that list as you wish using the fill handle in cell **A101**. The next column, labeled SDATA, is reserved to hold any sorted data set for which you wish to construct Nscores, and the third column, labeled NSCORES, contains the formulas for converting the sorted data into Nscores. For this example, proceed as follows.

Excel commands:

1. Click on the DATA worksheet tab and select the AGI data in the range **X2:X13**
2. Choose **E**dit ➤ **C**opy (or click the copy button in the toolbar) and click on the NSCORES worksheet tab
3. Select cell **B2**, choose **E**dit ➤ **P**aste (or click the paste button in the toolbar) and delete any data, if necessary, below the last data entry
4. Click the column button **B** and then click the [A↓Z] button in the toolbar

You will discover that the NSCORES column has now been filled in with the Nscore value corresponding to each of the sorted data points. The reason this can be automatically accomplished is that the Nscore value only depends on the rank of the ordered data point and the total number of data points. This is why we included the RANK column at the start and it is also the reason that you must exercise care to erase or delete all data below those that are copied in Step 3. If values are left in the cells below the last one copied, they will be included in the sorting.

Now we are prepared to plot the Nscore values against the AGI values using Excel's Chart Wizard by executing the following steps. Remain in the NSCORES worksheet. Recall that a *normal probability plot* is a plot the Nscore against the variable itself. So proceed to execute the steps on the following page.

76 Lesson 6 The Normal Distribution

Excel commands:

1. Click the chart wizard button 📊 in the toolbar, select **XY (Scatter)** as the **Chart type:**, then click [Next >]
2. Use the 📋 button in the **Data range:** text box to select the range **B2:Cx** where x is the row containing the last data value
3. Select the **Columns** radio button and click [Next >]
4. In the **Legend** folder, deselect **Show legend** check box and click [Next >]
5. In the **Titles** folder, type AGI in the **Value(X) Axis** text box
6. Type NSCORES in the **Value(Y) Axis** text box and click [Next >]
7. Select the **As object in:** radio button, click ▼ in its text box and select **NSCORES**
8. Select [Finish]

Excel creates a normal probability plot in the **NSCORES** worksheet. Move the chart so that it is clear of any data. We formatted the chart further as follows.

Excel commands:

1. Right click anywhere on the Value (Y) axis in the chart, select **Format Axis...**
2. Click the **Scale** tab if necessary and type −2 in the **Value (X) axis Crosses at:** text box

The result is shown in Figure 6.3.

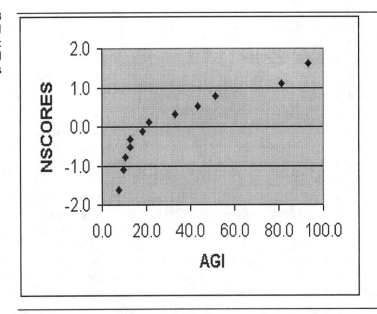

FIGURE 6.3 Excel normal probability plot for adjusted gross incomes

You will notice considerable curvature in the probability plot indicating that AGI values are not normally distributed. ∎

PROBLEM 6.6
The exam scores for the students in an introductory statistics course are as follows.

88	67	64	76	86
85	82	39	75	34
90	63	89	90	84
81	96	100	70	96

The data are stored in a column labeled **SCORES** in the **DATA** worksheet. Use Excel to construct a normal probability plot for these data, and make a judgment about the assumption that these data arose from a normal population. ◇

DETECTING OUTLIERS

E-364:I-388

Recall that outliers are data values that fall well outside the overall pattern of a data set. Normal probability plots can help us make judgments about the nature of such outliers, as the following example illustrates.

EXAMPLE 6.9 Using normal probability plots to detect outliers

The data on chicken consumption in the United States reported in Table 6.5 on page 387 of *IS/5e* (page 363 of *ES/4e*) are repeated here in Table 6.2.

TABLE 6.2 Sample of last year's chicken consumption (lbs)

47	39	62	49	50	70
59	53	55	0	65	63
53	51	50	72	45	

Use Excel to detect outliers in the data.

SOLUTION Following the instructions just given for constructing normal scores and plotting the results, we obtained the normal probability plot in Figure 6.4.

FIGURE 6.4 Excel normal probability plot for chicken consumption

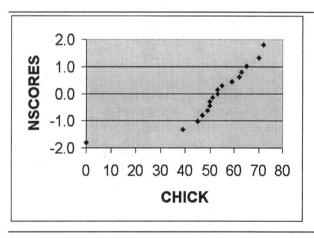

Notice that the data fall roughly in a straight line except for the value at 0. This would suggest that the subpopulation of people who consume no chicken at all are not representative of the population of chicken consumers. In other words, 0 appears to be an outlier so we remove it from the sample. When that is done, the probability plot then appears as in Figure 6.5.

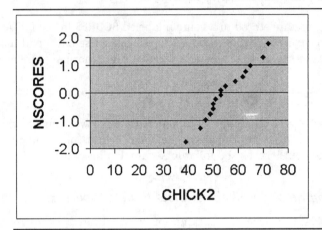

FIGURE 6.5
Excel normal probability plot for adjusted chicken consumption

The new sample is more reasonably linear. Hence, the assumption of normality appears to be safe to use for these data. Perhaps it would be appropriate also to adjust our conclusions to apply only to chicken consumers in the population as a whole. ∎

PROBLEM 6.7   The U.S. Federal Highway Administration conducts studies on motor vehicle travel by type of vehicle. Results are published annually in *Highway Statistics*. A sample of 15 cars yields the following data on number of miles driven, in thousands, for last year.

10.2	10.3	8.9	12.7	8.3
9.2	13.7	7.7	3.3	10.6
11.8	6.6	8.6	5.7	12.0

The data are stored a column labeled MILES in the DATA worksheet. Use Excel to analyze these data for outliers.   ◇

LESSON 7

The Sampling Distribution of the Mean

GENERAL OBJECTIVE Excel is also quite useful for amplifying and clarifying concepts related to sampling and sampling distributions. In this lesson we will employ Excel to sample from a population to illustrate sampling error, the sampling distribution of the mean and the central limit theorem.

LESSON OUTLINE
7.1 Sampling distributions; sampling error
7.2 The sampling distribution of the mean

7.1 Sampling distributions; sampling error

E-376 : I-408

A sample from a population provides us with only a portion of the entire population. Thus it is unreasonable to expect that the sample will yield perfectly accurate information about the population. Consequently, we should anticipate that a certain amount of error will result simply because we are sampling. This type of error is aptly called *sampling error*. We illustrate with an example.

EXAMPLE 7.1 Illustrates sampling error

Suppose that the *population* under consideration consists of the annual salaries of the top five state officials of Oklahoma. The population of salaries is presented in the second column of Table 7.1. [Data from *The World Almanac, 1998*.] Values are in thousands of dollars.

TABLE 7.1
Top five Oklahoma state officials and their annual salaries

Official	Salary ($1000s)
Governor (G)	70
Lieutenant Governor (L)	63
Secretary of State (S)	44
Attorney General (A)	75
Treasurer (T)	70

Use Excel to draw a sample of size 2 (without replacement) and compare the sample mean with the population mean.

SOLUTION We stored the population of five salaries (the second column of Table 2.1) in a column labeled SALARY in the Excel worksheet DATA of the Probability workbook.

Since we need to sample without replacement, we have to exercise a little caution in how we proceed because 70 occurs twice in the population. Unless we distinguish somehow between these two 70's we may upset the sampling scheme because we are restricted to sampling *with* replacement in Excel. (See Lesson 1.) Look at it this way. What we really aim to do is sample two distinct state officials and look at their salaries. Thus, in this scheme, it is possible to have 70 arise twice but if we sample with replacement, as we are bound to do in Excel, we would have no way of knowing if the two resulted from sampling the same official twice (hence we should break the tie) or from two distinct officials (hence we would not break the tie).

A simple way out of this dilemma is to choose two distinct officials first and then select their salaries. You could just use their official titles to sample from or, better still, we named a column labeled CODES and entered the integers 1 through 5 in cells AD2 through AD6. We will draw a sample of size 2 from the codes, without replacement, and then match the result to the salaries. We labeled a column SMPLC to hold the sample code values and another column SMPLS to hold the sample of salaries.

We then followed the Excel commands located in Section 1.5 of Lesson 1 which we repeat here.

Excel commands:

1. Choose **Tools** ➤ **Data Analysis**...
2. Select **Sampling** in the **Analysis Tools** option box
3. Use the ▥ button in the **Input Range:** text box, to select the range **AD2:AD6**
4. Select the **Random** radio button in the **Sampling Method** option box, select the **Number of Samples:** text box and type **2**
5. Click the **Output Range:** option button, and use the ▥ button in the text box to select cell **AE2**
6. Click [OK]

The two sample codes will appear in column **SMPLC** and the corresponding salaries will appear in column **SMPLS**. If they are the same, break the tie by repeating the sampling procedure on one of them (typing **1** instead of **2** at Step 4), and repeat if necessary until there are no ties.

We drew codes 1 and 3 the first time out so no tie break was necessary. The two salaries 70 and 44 were obtained. In cell **AF4** we typed `=AVERAGE(AF2:AF3)` to record the *sample* mean as 57.0. Thus, the sample mean of the two randomly selected salaries is given by $\bar{x} = \$57,000$.

In cell **AC7** we typed `=AVERAGE(AE2:AE6)` to see that the mean of this population is 64.4, or $64,400. Comparing the sample mean of $\bar{x} = 57.0$ to the population mean of $\mu = 64.4$, we see that in this case the sampling error made in estimating μ by $\bar{x}$ is 7.4 (that is, $7,400) in absolute terms. ∎

PROBLEM 7.1 Repeat Example 7.1 three times. *(Note:* It is possible that you may obtain the same random sample more than once.) ◇

7.2 The sampling distribution of the mean

An important tool in statistical analysis is related to knowledge of the probability distribution of the mean, $\overline{X}$. The probability distribution of this random variable is so important that it is given a special name—the *sampling distribution of the mean*.

The sampling distribution of the mean depends on the distribution of the population being sampled and on the sample size. In general, it is not possible to obtain the sampling distribution of the mean *exactly*. Even though it is generally not possible to determine the sampling distribution of the mean exactly, there are mathematical relationships that permit us to obtain the sampling distribution of the mean approximately. It is helpful to distinguish between the case where the population being sampled is normally distributed and the case where it may not be so. Let us first consider normally distributed populations.

NORMALLY DISTRIBUTED POPULATIONS

E-394 : I-426

Suppose the population under consideration is *normally distributed*. Then, although it may not be obvious, the random variable $\overline{X}$ is also normally distributed. Specifically we have the following result:

KEY FACT 7.1 **The sampling distribution of the mean for normal populations**

Suppose a random sample of size n is to be taken from a *normally distributed* population with mean μ and standard deviation σ. Then the random variable $\overline{X}$ is also *normally distributed* and has mean $\mu_{\overline{x}} = \mu$ and standard deviation $\sigma_{\overline{x}} = \sigma/\sqrt{n}$. In other words, if the population being sampled is normally distributed, then probabilities for $\overline{X}$ are equal to areas under the normal curve with parameters μ and $\sigma/\sqrt{n}$.

An intuitive interpretation of this Key Fact is as follows: Suppose we compute the sample mean, $\overline{x}$, for each possible sample of size n from a normally distributed population. Then the histogram of all of those $\overline{x}$-values will be bell-shaped. Let us now illustrate this intuitive interpretation with the aid of Excel.

EXAMPLE 7.2 *Illustrates the sampling distribution of the mean for normal populations*

The ages of farm operators in the United States are assumed to be *normally distributed* with a mean of $\mu = 50$ years and a standard deviation of $\sigma = 8$ years. Apply Excel to illustrate the sampling distribution of the mean for this population of ages when $n = 3$.

SOLUTION   Since the population is normally distributed, the random variable $\overline{X}$ is also normally distributed and has mean $\mu_{\overline{x}} = \mu = 50$, the same as the population mean. However, the standard deviation is $\sigma_{\overline{x}} = \sigma/\sqrt{n} = 8/\sqrt{3} = 4.62$. Thus the set of all possible $\overline{x}$-values obtained from all possible samples of size $n = 3$ has a normal distribution with mean $\mu = 50$ and standard deviation $\sigma = 4.62$.

Unfortunately, we cannot get all the $\overline{x}$-values since, at least theoretically, there are an infinite number of them. However, we can make plausible the fact that the set of all $\overline{x}$-values is normally distributed by obtaining a large number of samples of size three and showing that the histogram of their $\overline{x}$-values is roughly bell-shaped.

In Lesson 6 you learned how Excel can be used to simulate random sampling from a normally distributed population. For our population of ages, we know that $\mu = 50$ and $\sigma = 8$. We employ Excel to (1) take 1000 samples of size $n = 3$, (2) compute $\overline{x}$ for each of the 1000 samples, and (3) draw a histogram of the 1000 $\overline{x}$-values. The resulting histogram should be roughly bell-shaped.

We now proceed to accomplish (1)–(3) using the **NSCORES** worksheet of the Probability workbook. First, we formatted columns K through N as numbers with 1

7.2 The sampling distribution of the mean

decimal place. Then we used Excel to take 1000 samples of size $n = 3$ from the population of ages as follows.

Excel commands:

1. Choose **Tools ▶ Data Analysis...**
2. Select **Random Number Generation** from the **Analysis Tools** option box
3. Select the **Number of Variables:** text box and type 3
4. Select the **Number of Random Numbers:** text box and type 1000
5. Click the ▼ button in the **Distribution:** drop down menu and select **Normal**
6. In the **Parameters** input box select the **Mean=** text box and type 50, select the **Standard Deviation=** text box and type 8
7. Select the **Output Range:** radio button in the **Output options** box and enter **K2** in its text box
8. Click [OK]

Each row of columns K, L, and M represents a single random sample of size 3. By averaging these rows, we can obtain 1,000 sample averages from this normal distribution. Continue as follows.

Excel commands:

1. Select cell **N2** and type `=AVERAGE(K2:M2)`
2. Click the copy button on the Excel toolbar
3. Select cell **N3**, press and hold the [Shift] key then press and hold down the [PgDn] key to scroll quickly through to cell **N1001**
4. Release both keys and click the paste button in the Excel toolbar

Now the 1,000 $\bar{x}$-values are located in column N (labeled xbars). In the range **O2:P6** you will find numbers to help decide on interval widths for the histogram. We decided on 2. We selected 36.0 for the first lower cutpoint and then proceeded by 2's as follows.

Excel commands:

1. Select input cell **P7** and type 2
2. Select input cell **P9** and type 36, then type 38 in cell **P10**
3. Select the range **P9:P10** and drag the fill handle of **P10** down until you see the limit of 66; that would be cell **P24**

We now have the necessary elements for the histogram. Notice that the range **R2:R3** contains the average of the xbars (49.9 here compared with $\mu = 50$) and the standard deviation of the 1,000 xbars (4.7 compared with $\sigma/\sqrt{3} = 4.62$). Proceed as follows to construct and display the histogram.

Excel commands:

1. Choose **Tools ▶ Data Analysis...** and select **Histogram** from the **Analysis Tools** option box

2 Type <u>xbars</u> in the <u>I</u>nput Range: text box

3 Use the button in the <u>B</u>in Range: text box to select the range **P9:P24**

4 Select the <u>O</u>utput Range: option button from the **Output Options** options box, type **Q8** in its text box and select the <u>C</u>hart output check box

5 Click [OK]

As before, edit the histogram by formatting the Data Series to have gap width 0, and rename the Category Axis Title **xbars**. Your histogram should resemble the one shown in Figure 7.1.

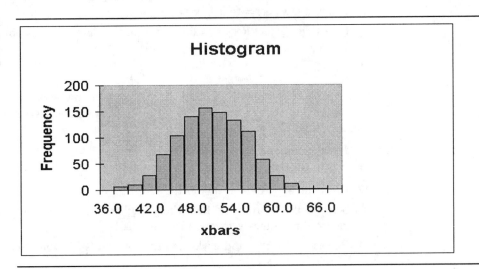

FIGURE 7.1 Sampling distribution of the mean; normal population

As we expected, the histogram of the 1000 $\bar{x}$-values is roughly bell-shaped. Your display may differ from this one because of the particular random samples you obtained. If we took, say, 5000 samples of size $n = 3$ and obtained a histogram of the 5000 $\bar{x}$-values, then that histogram would be even more bell-shaped. ∎

PROBLEM 7.2 Refer to Example 7.2. Use Excel to
a) take your own 1000 random samples of size $n = 3$ from the population of ages.
b) compute $\bar{x}$ for each sample obtained in part (a).
c) draw a histogram of the 500 $\bar{x}$-values from part(b). ◇

CENTRAL LIMIT THEOREM [E-397 : I-429]

We have seen that if the population being sampled is normally distributed, then so is the sampling distribution of the mean (that is, the probability distribution of the sample mean, $\overline{X}$). Remarkably, when the sample size is relatively large, the sampling distribution of the mean is approximately a normal distribution, even if the population being sampled is not normally distributed. This extraordinary fact is called the *central limit theorem*:

KEY FACT 7.2 The central limit theorem

For a relatively large sample size, the random variable $\overline{X}$ is approximately *normally distributed*, regardless of how the population is distributed. The approximation becomes better and better with increasing sample size.

In general, the more "nonnormal" the population, the larger the sample size must be in order to use normal curve areas to determine probabilities for $\overline{X}$. As a rule of thumb, we will consider sample sizes of 30 or more ($n \geq 30$) large enough. The following statement summarizes the result.

KEY FACT 7.3 The sampling distribution of the mean for general populations

Suppose a random sample of size $n \geq 30$ is to be taken from a population with mean μ and standard deviation σ. Then regardless of the distribution of the population, the random variable $\overline{X}$ is approximately *normally distributed* and has mean $\mu_{\overline{x}} = \mu$ and standard deviation $\sigma_{\overline{x}} = \sigma/\sqrt{n}$. That is, probabilities for $\overline{X}$ are approximately equal to areas under the normal curve with parameters μ and $\sigma/\sqrt{n}$.

In the next example we will see how Excel can be used to illustrate the central limit theorem and the sampling distribution of the mean for a population that is *definitely not* normally distributed.

EXAMPLE 7.3 *Illustrates the central limit theorem and sampling distribution of the mean for nonnormal populations*

A telephone company in the Southwest undertook a study of telephone usage in Phoenix, Arizona. The results indicate that the durations of telephone conversations by residential customers in Phoenix have an *exponential distribution* with a parameter μ roughly equal to 4 minutes. This means the following: Let the population under consideration be the durations, in minutes, of all possible telephone conversations by Phoenix residential customers. Then percentages for that population are equal to areas under the curve

$$y = \frac{1}{4}e^{-x/4}.$$

This curve is pictured in Figure 7.2 on the following page. Clearly this population is not normally distributed. Nonetheless, according to the central limit theorem, the random variable $\overline{X}$ is approximately normally distributed when the sample size is relatively large (30 or more).

FIGURE 7.2
Exponential curve for duration of conversation

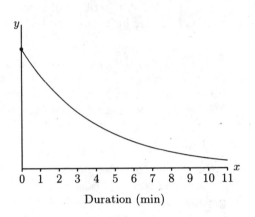

Duration (min)

Use Excel to illustrate this fact, using 250 samples of sample size 30 for each sample.

SOLUTION We first explain how Excel can be used to simulate a random sample from an exponential distribution. Unfortunately, there is no Exponential choice for the **Distribution**: text box in the **Random Number Generation** Analysis Tool. However, there is an elegant theorem that allows us to match an output from a Uniform Distribution (Exercise 16 on page 438 of *IS/5e* (page 406 of *ES/4e*). This distribution over the interval from 0 to 1, abbreviated (0, 1), is analogous to the equally likely probability model only on a continuous scale. That is, you are just as likely to find yourself in one part of the interval as another when you draw a sample *at random*.

Hence we will first draw 250 random samples of size 30 each from the uniform distribution. Proceed as follows, using our choices in the **NSCORES** worksheet. (Of course, you may alter these if you wish, perhaps using a different worksheet.)

Excel commands:

1. Choose **T**ools ➤ **D**ata Analysis...
2. Select **Random Number Generation** from the **A**nalysis **T**ools option box
3. Select the **Number of V**ariables: text box and type 30
4. Select the **Number of Random Numbers:** text box and type 250
5. Click the ▼ button in the **D**istribution: drop down menu and select **Uniform**, accepting the default parameters of **0** and **1**
6. Select **O**utput Range: in the **Output options** box and enter **AA252** in its text box
7. Click [OK]

7.2 The sampling distribution of the mean

Each row in the range **AA252:BD501** represents a single random sample of size 30 from the uniform distribution. To convert these to the exponential distribution having a mean of 4 (minutes) proceed as follows.

Excel commands:

1 Select cell **AA2**, type =-LN(AA252) and press [Enter↵]
2 Click the copy button 🗐 on the Toolbar or choose **Edit ➤ Copy**
3 Select the range **AA2:BD251**
4 Click the paste button 🗐 on the Toolbar or choose **Edit ➤ Paste**

Each row of the range **AA2:BD251** now represents a single random sample of size 30 from the exponential distribution having a mean of 4 min. By averaging these rows, we can obtain 250 sample averages from this exponential distribution. Continue as follows.

Excel commands:

1 Select cell **BE2** and type =AVERAGE(AA2:BD2)
2 Drag the fill handle of cell **BE2** down through cell **BE251**

Now the 250 $\bar{x}$-values are in column **BE**, labeled **xbar**. To ease construction of the histogram, copy the range **O2:P10** and paste it into cell **BF2**; copy the range **Q2:R3** and paste it into cell **BH2**. In the range **BF2:BG6** you will find numbers to help decide on interval widths for the histogram. We decided on 0.5. We selected 2.0 for the first lower cutpoint and then proceeded by 0.5's as follows.

Excel commands:

1 Select cell **BG7** and type 0.5
2 Select cell **BG9** and type 2.0 then type 2.5 in cell **BG10**
3 Select the range **BG9:BG10** and drag the fill handle of **BG10** down until you see the limit of 7; that would be cell **BG19**

We now have the necessary elements for the histogram. Of course, each implementation may lead to different values, hence slightly different locations. Notice also that the range **BH2:BI3** contains the average of the xbars (3.9 here compared with $\mu = 4$). Proceed as follows to construct and display the histogram.

Excel commands:

1 Choose **Tools ➤ Data Analysis...** and select **Histogram** from the **Analysis Tools** option box
2 Select the **Input Range:** text box and type xbar
3 Use the 🔼 button in the **Bin Range:** text box to select the range BG9:BG19
4 Select the **Output Range:** radio button from the **Output Options** options box, type **BH8** in its text box and select **Chart output**
5 Click [OK]

As before, edit the histogram by formatting the Data Series to have gap width 0, and rename the Category Axis Title **xbars**. Your histogram should resemble the one shown in Figure 7.3.

FIGURE 7.3
Sampling distribution of the mean; exponential population

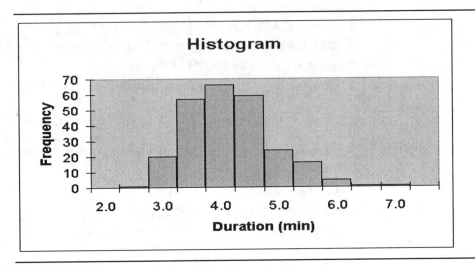

The histogram in Figure 7.3 seems to confirm that the sampling distribution is approximately normally distributed while the population is not normally distributed. Even though the distribution of xbar's is still slightly skewed, it bears little resemblance to the population distribution curve of Figure 7.2. Selecting a larger sample size would make the **normal** appearance even more dramatic. ■

PROBLEM 7.3 Refer to Example 7.3 on page 85.
a) Use Excel to (1) take 500 samples of size $n = 2$ from the population of phone conversation durations, (2) compute $\bar{x}$ for each of the 500 samples obtained, and (3) draw a histogram of the 500 $\bar{x}$-values. Interpret your results in light of the central limit theorem.
b) Repeat part (a) for samples of size $n = 10$. It is recommended that you use a separate worksheet using the one provided here as a model. ◊

LESSON 8

Confidence Intervals for One Population Mean

GENERAL OBJECTIVE

We now begin our study of how Excel can be used in inferential statistics. In this lesson, we will employ Excel to obtain *confidence intervals* for a population mean, μ. Specifically, we will learn the Excel functions that permit us to perform Procedures 8.1 and 8.2 in *IS/5e* and *ES/4e* by computer.

Excel has a very limited capability for computing confidence intervals from a given set of data, much less from summary data, directly. However, the program does contain more than enough of the required functions to construct a permanent template for carrying out all of the procedures in this and the next few chapters.

Consequently, we have constructed an Excel workbook called **Inference** which you will find on your distribution disk. This notebook contains several worksheets that automate the process of finding confidence intervals and testing hypotheses (the next chapter) in the common situations encountered in statistics. In addition, we have registered and named all of the data sets used in the book in a separate worksheet called **DATA** which is coordinated with the other worksheets. This one workbook will implement all of the procedures in Chapters 8–12 in *IS/5e* and *ES/4e*, as well as Chapter 13 of *IS/5e*.

Once again we caution you to preserve an original copy of the Excel files distributed to you with this manual. If some of the crucial cells happen to be changed by accident, there could be irreparable damage to the output.[†] You must have an original workbook to copy into your computer in that case.

LESSON OUTLINE

8.1 Confidence intervals for one population mean; σ known
8.2 Confidence intervals for one population mean; σ unknown

[†] That is one of the reasons that we recommend you prevent the cursor from moving automatically after a cell has been entered (see Lesson 1, page 8).

8.1 Confidence intervals for one population mean; σ known

E-422:I-454

Procedure 8.1 on page 452 of *IS/5e* (page 420 of *ES/4e*) gives a step-by-step method for obtaining a confidence interval for a population mean, μ, when σ is known. We will implement this procedure with an example.

Example 8.4 on page 454 of *IS/5e* (page 422 of *ES/4e*) illustrates the application of Procedure 8.1 to determine a confidence interval for the mean age of all people in the civilian labor force. Let us use the **Inference** workbook functions to obtain that confidence interval.

EXAMPLE 8.1 *Illustrates computing a z-confidence interval*

The U.S. Bureau of Labor Statistics collects and publishes data on the ages of people in the civilian labor force. Suppose $n = 50$ such people are randomly selected and that their ages are as displayed in Table 8.1. Assume $\sigma = 12.1$ years.

TABLE 8.1
Ages of 50 randomly selected people in the civilian labor force

22	58	40	42	43	32	34	45	38	19
33	16	49	29	30	43	37	19	21	62
60	41	28	35	37	51	37	65	57	26
27	31	33	24	34	28	39	43	26	38
42	40	31	34	38	35	29	33	32	33

Use Excel to find a 95% confidence interval for the mean age, μ, of all people in the civilian labor force.

SOLUTION We have entered the sample data of Table 8.1 into a column labeled **AGES** in the **DATA** worksheet of the **Inference** workbook. Open the **1SAMP** worksheet, find the panel labeled **One Mean**, and proceed as follows.

Excel commands:

1 Type <u>AGES</u> in input cell **B4** under **DATA FILE NAME:** and press ⏎
2 Type <u>12.1</u> in input cell **E3** for the value of σ and press ⏎
3 Type <u>0</u> for **Alt Code** in input cell **D9** (see next chapter)
4 Type <u>.05</u> in input cell **D10** for the value of α and press ⏎

Figure 8.1 on the following page shows the results. Ignoring all other outputs, you will find the results recorded in the range **E11:F13** under the heading **z-Confidence Interval** in the worksheet. Thus a 95% confidence interval for μ is from 33.026 to 39.734. We can be 95% confident that the mean age, μ, of all people in the civilian labor force is somewhere between 33.026 and 39.734 years. ■

8.1 Confidence intervals for one population mean; σ known

FIGURE 8.1
Excel output
for One Mean
using AGES data

One Mean				
DATA FILE NAME:		$\sigma =$	12.100	
AGES	n =	50	df =	
	xbar =	36.380	Min =	16.00
	s =	11.069	$Q_1 =$	29.25
	SE Mean	1.71	Median =	34.5
	$H_0: \mu_0 =$	0.00	$Q_2 =$	41.75
	Alt Code =	0	Max =	65.00
z-Test:	$\alpha =$	0.05	z-Confidence	Interval:
Two-tailed test: z =	21.260		95%	Two-sided:
Lower Critical Value:	-1.960		Lower Limit l =	33.026
Upper Critical Value:	1.960		Upper Limit u =	39.734
P-value:	0.0000		$\mu_0 =$	0.00
Action:	Reject		Action:	Reject

Before proceeding, there are some general remarks that we would offer for the **One Mean** panel. First, for each procedure, cells that require an input from you (called *input cells*) have a dotted border as previously noted, although not every such cell requires an input for every single procedure; it is reasonably safe to ignore the ones that do not require input.

Secondly, the range **C4:F9**, displays a set of descriptive statistics for you once you select the data set. This includes the five-number summary (see Lesson 3) in case you have need for further analysis of the problem situation.

Finally, the panel is arranged to give you results dynamically. You have merely to replace the name of the data set by another name, or range preceded by location (like **DATA!G2:G13**), if the data set is not named. The results of the new data set are given immediately. Indeed, it is only necessary to have summary data to compute these results. Sometimes, you will not have access to the original data set but will have a published set of summary statistics. In that case, proceed as in the next example.

EXAMPLE 8.2   *Illustrates computing a confidence interval using only summary data*

In the circumstances of the last example, suppose that you only know the summary statistics $n = 50$, $\bar{x} = 36.38$. Assume $\sigma = 12.1$ years as before. Use Excel to find a 95% confidence interval for the mean age, μ, of all people in the civilian labor force.

SOLUTION   Open the **DATA** worksheet. Column **B** has been given the name **SD** (for *summary data*). Proceed as follows.

Excel commands:

1 Type 50 in input cell **B2** next to n ==> for the sample size n and press [Enter↵]
2 Type 36.38 in input cell **B4** next to xbar ==> and press [Enter↵]

3 Click on the **1SAMP** tab and locate the panel **One Mean**
4 Type SD in input cell **B4** under the heading **DATA FILE NAME:** and press [Enter⏎]
5 Type 12.1 in input cell **E3** for the value of σ and press [Enter⏎]
6 Type 0 for **Alt Code** in input cell **D9**, type .05 in cell **D10** for the value of α and press [Enter⏎]

The output in the range **E10:F13** is precisely the same as before. Of course, without the original data, we cannot compute summary statistics. ∎

PROBLEM 8.1   The U.S. National Center for Health Statistics estimates mean weights and heights of U.S. adults by age and sex. Suppose that 40 women, 5 feet 4 inches tall and aged 18–24, are randomly selected and that their weights, in pounds, are as shown here.

140	136	147	138	143	122	115	125
136	152	130	134	150	153	148	132
116	159	128	136	134	126	120	146
131	167	145	132	138	137	115	145
154	139	139	147	123	154	127	116

Use Excel to find a 90% confidence interval for the mean weight, μ, of all U.S. women 5 feet 4 inches tall and in the age group 18–24. The data are stored in a column labeled **WEIGHT**. Assume $\sigma = 12.8$. ◇

8.2 Confidence intervals for one population mean; σ unknown

We can also use Excel to obtain a confidence interval for a population mean, μ, when the population being sampled is normally distributed and the standard deviation σ is unknown. The steps are summarized as Procedure 8.2 on page 471 of *IS/5e* (page 439 of *ES/4e*). The formulas involve finding *t*-values corresponding to a specified area under a *t*-curve, that is, percentile points and we investigate that problem first.

FINDING THE t-VALUE FOR A SPECIFIED AREA   E-438 : I-470

Excel has a function called **TINV**, which has the format TINV(α,df) and returns the value $t_{\alpha/2}$, provided $\alpha < \frac{1}{2}$. Here, df is a specification of degrees of freedom and $\alpha/2$ is the desired area to the *right*. You need to supply α and df as arguments for the function.

EXAMPLE 8.3 Illustrates finding the t-value for a specified area using TINV

For a t-curve with 13 degrees of freedom, find the t-value with area 0.05 to its left using Excel.

SOLUTION We seek the value of $t_{.95}$. Using Property 3 discussed on page 469 of *IS/5e* (page 437 in *ES/4e*), we know that $t_{.95} = -t_{.05}$. Hence, choose $\alpha = 0.10$ for input. Find any empty cell in any worksheet and type `=TINV(.10,13)`. The cell contents will be replaced instantly by 1.770932. This is $t_{.05}$ and hence $t_{.95} = -1.77$. ∎

PROBLEM 8.2

For a t-curve with df=8, use Excel to find the following t-values.
a) The t-value with area 0.10 to its right.
b) $t_{0.005}$.
c) The t-value with area 0.01 to its left. ◇

CONFIDENCE INTERVALS: t-INTERVAL [E-441 : I-4473]

A step-by-step method of finding a confidence interval for the mean when σ is unknown is outlined in Procedure 8.2 on page 442 of *IS/5e* (page 439 of *ES/4e*). The following example shows how to determine that confidence interval by employing Excel.

EXAMPLE 8.4 Illustrates Procedure 8.2 for computing a t-interval

The U.S. Federal Bureau of Investigation (FBI) compiles data on robbery and property crimes and publishes the information in *Population-at-Risk Rates and Selected Crime Indicators*. A sample of last year's pick-pocket offenses yields the values lost shown in Table 8.2.

TABLE 8.2
Value lost ($) for a sample of 25 pick-pocket offenses

447	207	627	430	883
313	844	253	397	214
217	768	1064	26	587
833	277	805	653	549
649	554	570	223	443

Use Excel to obtain a 95% confidence interval for the mean value lost, μ, of all last year's pick-pocket offenses.

SOLUTION We need to be reasonably sure that the population of values lost is normally distributed if we are to apply the methods of this section, for the sample size is not large. We constructed the probability plot shown in Figure 8.2 on the following page using the methods of Lesson 6.

FIGURE 8.2
Excel normal probability plot of VALUE data

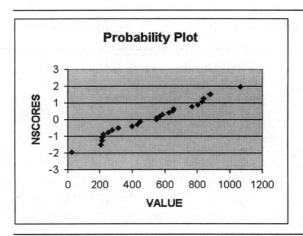

First, we stored the sample data of Table 8.2 into a column named VALUE in the **DATA** worksheet, then copied the data to a column named SVALUE in the **CHARTS** worksheet, where the data were sorted. Next, we converted the sorted data to Nscores then plotted the Nscores against the data using the XY (Scatter) option of the Chart Wizard (as in Example 6.8, page 75).

The normal probability plot in Figure 8.2 shows no outliers and falls roughly in a straight line. The value at 26 is a little worrisome, not so much in terms of linearity, with the rest of the points as the possibility of being an outlier. Even a rough look at the data would suggest this before any statistical analysis. However, the violations do not appear to be too extreme.

That being the case, we proceeded as follows to find a 95% confidence interval using the Inference notebook in Excel.

Excel commands:

1 Click on the **1SAMP** worksheet tab and locate the **One Mean** panel
2 Type VALUE in input cell **B4** under **DATA FILE NAME:** and press Enter⏎
3 Clear (delete) any entry in input cell **E3** for the value of σ
4 Type 0 for **Alt Code** in input cell **D9** and type .05 in input cell **D10** for the value of α

Ignoring all other outputs, you will find the results recorded in the range **E10:F13** under the heading **t-Confidence Interval** in the worksheet. Figure 8.3 shows the output for this range.

FIGURE 8.3
Excel output for **One Mean** using VALUE data

t-Confidence	Interval:
95%	Two-sided:
Lower Limit l =	405.076
Upper Limit u =	621.564

We can be 95% confident that the mean value lost, is somewhere between $405.08 and $621.56. ∎

Incidentally, for this last example, you may check the range **D4:D6** to discover that $n = 50$, $\bar{x} = 513.32$, and $s = 262.23$. This is all the information that is needed to complete this test. Try entering these summary data in the **DATA** worksheet, select SD for the **DATA FILE NAME:** in the **1SAMP** and you will discover the same output shown in Figure 8.3.

PROBLEM 8.3 According to the R.R. Bowker Company of New York, the mean annual subscription rate to law periodicals was $97.33 in 1995. In 1996, a random sample of $n = 12$ law periodicals yields current annual subscription rates, to the nearest dollar, as displayed in the following table.

106	122	120	123
118	114	138	131
128	124	119	130

Test the 1996 annual subscription rates to law periodicals for normality. If you are satisfied that rates are normally distributed, use Excel to obtain a 95% confidence interval for this year's mean annual subscription rate, μ, for law periodicals. The data are stored in a column named **SUBRATES**. ⋄

LESSON 9

Hypothesis Tests for One Population Mean

GENERAL OBJECTIVE In Lesson 8 we discovered how Excel can be used to obtain confidence intervals for one population mean, μ. Now we will examine the Excel functions that perform *hypothesis tests* for μ. These functions correspond to various Procedures in *IS/5e* and *ES/4e*.

LESSON OUTLINE
9.1 Some preliminaries
9.2 One-sample z-test for a population mean
9.3 One-sample t-test for a population mean

9.1 Some preliminaries

Before commencing our study of how Excel can be used to perform hypothesis tests for one population mean, μ, we need to consider some preliminaries. We begin by reviewing how the alternative hypothesis is specified.

SPECIFYING THE ALTERNATIVE HYPOTHESIS

Recall that typically there are two hypotheses in a hypothesis test—the **null hypothesis** (H_0) and the **alternative hypothesis** (H_a). The problem in a hypothesis test is to decide whether or not the null hypothesis should be rejected in favor of the alternative hypothesis.

In hypothesis tests concerning a single population mean, μ, the null hypothesis generally specifies a *single value* for that parameter. In other words, the null hypothesis is of the form

$$H_0 : \mu = \mu_0,$$

where μ_0 is some specified number.

On the other hand, there are *three* possibilities for the alternative hypothesis. It can be of the form $H_a : \mu \neq \mu_0$, $H_a : \mu < \mu_0$, or $H_a : \mu > \mu_0$. The hypothesis test is called **two-tailed** in the first case, **left-tailed** in the second case, and **right-tailed** in the third case. Table 9.1 summarizes this discussion.

TABLE 9.1 Possible alternative hypotheses for hypothesis tests concerning a single population mean

Type	Form
Two-tailed	$H_a : \mu \neq \mu_0$
Left-tailed	$H_a : \mu < \mu_0$
Right-tailed	$H_a : \mu > \mu_0$

We have organized the **1SAMP** worksheet in the **Inference** workbook to execute and display the results for any of the three alternatives. You have only to select the appropriate one to apply to a given situation and enter it in the worksheet as an alternative code, **Alt Code**, as follows.

To perform a left-tailed test, type -1 (or any negative number) in the input cell; to perform a right-tailed test, type 1 (or any positive number); to perform a two-tailed test, type 0. See Table 9.2.

TABLE 9.2 Codes for specifying the alternative hypothesis in Excel

Type of test	Subcommand
Two-tailed	Alt Code = 0
Left-tailed	Alt Code = -1
Right-tailed	Alt Code = 1

P-VALUES

Next we discuss P-values. The **P-value** for a hypothesis test is the smallest significance level at which the null hypothesis can be rejected with the observed sample data. When we use Excel to perform a hypothesis test, the P-value is almost always printed out. To test a hypothesis using the P-value, we have the following decision criterion.

KEY FACT 9.1 Decision criterion for a hypothesis test using the P-value
If the P-value is less than or equal to the specified significance level, α, then reject the null hypothesis. Otherwise, do not reject the null hypothesis.

You will get a lot of practice with P-values throughout the remainder of this manual. In the **1SAMP** worksheet you will find the P-value computed for everyone of the procedures we study in this book.

RELATIONSHIP BETWEEN HYPOTHESIS TESTS AND CONFIDENCE INTERVALS

The equivalence of testing a hypothesis and observing whether or not the appropriate confidence interval contains the hypothesized value is discussed briefly on page 521 of *IS/5e* (page 489 of *ES/4e*). The details are left as an exercise, Exercise 9.53 on page 524 in *IS/5e* and page 492 in *ES/4e*. The discussion there is restricted to a two-tailed test, however. With some modification, the same equivalence may be found for one-tailed tests as well.

Let us examine the Right-tailed test discussed in Procedure 9.1 on page 513 of *IS/5e* (page 481 of *ES/4e*). You see there that you *FAIL* to reject the null hypothesis when

$$\frac{\overline{x} - \mu_0}{\sigma/\sqrt{n}} < z_\alpha$$

and, otherwise, you reject the null hypothesis. But it is easy to verify that this inequality is equivalent to stating

$$\overline{x} - z_\alpha \cdot \sigma/\sqrt{n} < \mu_0.$$

However, if you will examine the formulas in Procedure 8.1 on page 452 of *IS/5e* (page 420 of *ES/4e*), you will see that the confidence limits for a confidence interval for μ with a confidence level of $1 - 2\alpha$ would be from

$$\overline{x} - z_\alpha \cdot \sigma/\sqrt{n} \quad \text{to} \quad \overline{x} + z_\alpha \cdot \sigma/\sqrt{n}$$

We call the interval running from

$$\overline{x} - z_\alpha \cdot \sigma/\sqrt{n} \quad \text{to} \quad \infty$$

a *lower one-sided confidence interval for* μ with confidence level $1 - \alpha$. Similarly, we call the interval running from

$$-\infty \quad \text{to} \quad \bar{x} + z_\alpha \cdot \sigma/\sqrt{n}$$

an *upper one-sided confidence interval for* μ with confidence level $1 - \alpha$.

To summarize, for the Right-tailed test you would *FAIL* to reject the null hypothesis when the value of μ_0 falls *INSIDE* the *LOWER* one-sided confidence interval; otherwise reject the null hypothesis (namely, when μ_0 is *OUTSIDE* the interval). Similarly, for the Left-tailed test you would *FAIL* to reject the null hypothesis when the value of μ_0 falls *INSIDE* the *UPPER* one-sided confidence interval; otherwise reject the null hypothesis (namely, when μ_0 is *OUTSIDE* the interval).

Notice the match Right-tail test to Lower one-sided interval and Left-tail test to Upper one-sided interval. It happens that this equivalence is valid in practically all of the procedures in this book when both hypothesis tests and confidence intervals are available. We have arranged the **1SAMP** worksheet so that the output displays the appropriate confidence interval to the side of each test. Not only that, the decision to reject or not reject H_0 is displayed as well. We will bring these matters to your attention in the examples.

Hypothesis testing and confidence interval estimation, as just outlined, are equivalent in the sense that both lead to exactly the same action.[†] That is, if one procedure leads to the final action of rejecting (or failing to reject) the hypothesis, the other will result in the same action. Yet the two methods have a slightly different perspective in terms of application.

Thus, in hypothesis testing, *for a given level of significance* α, the P-value tells at a glance, all significance levels at which the hypothesis would be rejected (those smaller than α) and those levels at which you would not reject the hypothesis (those greater than α).

On the other hand, *for a given level of significance* α, the confidence interval reveals at a glance all those choices for H_0 for which the hypothesis would be rejected (those outside the confidence interval) and those choices for which you would not reject the hypothesis (those inside the confidence interval).

Another view of this difference is that if you change the level of significance, α, the P-value does not change but the confidence interval does. If you change the hypothesis, H_0, you wish to test, the confidence interval does not change but the P-value does.

9.2 One-sample z-test for a population mean E-499 : I-542

We can use Excel to perform the one-sample z-test (Procedure 9.2 on page 541 of *IS/5e*, page 498 of *ES/4e*), a hypothesis test for a population mean, μ, when the standard deviation, σ, is known. At the same time, we can illustrate the P-value approach and the equivalence to confidence intervals. To explain the details, we will use Example 9.14 on page 540 of *IS/5e* (Example 9.12 on page 497 of *ES/4e*).

[†] For an exception, see Problem 12.3.

EXAMPLE 9.1 Illustrates the one-sample z-test

Calcium is the most abundant and one of the most important minerals in the body. It works with phosphorus to build and maintain bones and teeth. According to the Food and Nutrition Board of the National Academy of Sciences, the recommended daily allowance (RDA) of calcium for adults is 800 mg (milligrams).

A nutritionist claims that the average person with an income below the poverty level gets less than the RDA of 800 mg. To test her claim, she obtains the daily intakes of calcium for a random sample of 35 people with incomes below the poverty level. The results are displayed in Table 9.3. Data are in milligrams.

TABLE 9.3 Daily calcium intakes

686	433	743	647	734	641
993	620	574	634	850	858
992	775	1113	672	879	609

At the 5% significance level, do the data provide sufficient evidence to conclude that the mean calcium intake of all people with incomes below the poverty level is less than the RDA of 800 mg per day? Use Excel to perform the hypothesis test. Assume $\sigma = 188$ mg.

SOLUTION Let μ denote the mean daily intake of calcium of all people with incomes below the poverty level. The problem is to perform the hypothesis test

H_0: $\mu = 800$ mg (mean calcium intake is not less than the RDA)

H_a: $\mu < 800$ mg (mean calcium intake is less than the RDA)

at the 5% significance level ($\alpha = 0.05$). Note that the hypothesis test is left-tailed since there is a less-than sign ($<$) in the alternative hypothesis.

We have entered the sample data from Table 9.3 into a column labeled **CALCIUM** in the **DATA** worksheet. Open the **1SAMP** worksheet, find the **One Mean** panel and proceed as follows.

Excel commands:

1. Select input cell **B4** under **DATA FILE NAME:** and type CALCIUM and press [Enter↵]
2. Select input cell **E3** and type 188 for the value of σ and press [Enter↵]
3. Select input cell **D8**, type 800 for the value of μ_0 and press [Enter↵]
4. Type -1 for **Alt Code** in input cell **D9** and press [Enter↵]
5. Type .05 in input cell **D10** for the value of α and press [Enter↵]

The results are displayed in the cell range **B10:F15** instantly as shown in Figure 9.1 on the following page. In addition, summary statistics for the data, including the five-number summary, are given.

As you can see, all three methods of testing can be employed at a glance. The fastest and easiest, is to observe that the computed P-value of 0.1176 is greater than the level of significance $\alpha = .05$. For a classical test, the test statistic has the value $z = -1.187$ and this is greater than the critical value of $-z_{.05} = -1.645$.

9.2 One-sample z-test for a population mean

FIGURE 9.1
Excel left-tailed test of CALCIUM data

	One Mean		
DATA FILE NAME:		$\sigma =$ 188.000	
CALCIUM	n = 18	df =	
	xbar = 747.389	Min =	433.00
	s = 171.989	Q_1 =	635.75
	SE Mean 44.31	Median =	710
	H_0: μ_0 = 800.00	Q_2 =	856
	Alt Code = -1	Max =	1113.00
z-Test:	α = 0.05	z-Confidence	Interval:
Left-tailed test: z = -1.187		95%	One-sided:
Critical Value: -1.645			
		Upper Limit u =	820.276
P-value: 0.1176		μ_0 =	800.00
Action: Do not Reject		Action:	Do Not Reject

Finally, the hypothesized value of $\mu_0 = 800$ is less than the upper limit of $u = 820.276$, that is, lies *inside* the upper one-sided confidence interval.

By all three criteria, you cannot reject the null hypothesis and you are reminded of this in cells **C15** and **F15**. The data do not provide sufficient evidence to conclude that the mean calcium intake of all people with incomes below the poverty level is less than the RDA of 800 mg per day. ∎

To solve a new problem, you have only to enter a new set of parameters and the output will be changed dynamically. To review, we call each cell requiring input from you an *input cell*; each is shown in the panel framed by dotted lines. There are five of these for a given scenario. The names of these input cells are given as follows: **DATA FILE NAME:**, $\sigma =$, $H_0 : \mu_0 =$, **Alt Code=**, and $\alpha =$.

PROBLEM 9.1 *The World Almanac, 1985* reports that the average travel time to work in 1980 for residents of South Dakota was 13 minutes. For this year, a random sample of 35 travel times for South Dakota residents yielded the data below (in minutes).

29	40	0	12	10	6	41
25	21	5	4	19	2	7
10	8	3	6	52	4	12
0	33	6	2	17	21	8
38	2	13	8	14	11	2

Use Excel to test, at the 5% significance level, whether this year's mean travel time to work for South Dakota residents appears to have *changed* from the 1980 mean of 13 minutes. Assume $\sigma = 11.6$ minutes. The data are stored and named **TRAVTIME** in the **DATA** worksheet. ◊

9.3 One-sample t-test for a population mean [E-509:I-552]

Procedure 9.3 beginning on page 550 of *IS/5e* (page 507 of *ES/4e*) is a step-by-step method for performing a hypothesis test concerning a population mean, μ, when the population being sampled is normally distributed or the sample size is large. We can use Excel to carry out such a hypothesis test just as easily as the z-test. The next example explains in detail how to apply this procedure.

EXAMPLE 9.2 *Illustrates the t-test for a mean*

The U.S. Energy Information Administration surveys households to obtain data on residential energy consumption and expenditures. Results of the surveys can be found in *Residential Energy Consumption Survey: Consumption and Expenditures*. According to that publication, the mean residential energy expenditure of all American families was $1282 in 1993. That same year, 36 randomly selected upper-income families reported the energy expenditures shown in Table 9.4.

TABLE 9.4 Energy expenditures ($) for 36 households living in single-family detached homes

2016	1509	1658	1359	1564	1808	1155	1948	1162
1529	956	1284	2224	1549	1751	1408	1124	1083
1406	1370	1719	1647	1151	1735	1421	1403	1134
1877	1296	1734	1275	1341	932	1564	1012	1309

At the 5% significance level, do the data indicate that, in 1993, upper-income families spent more, on the average, for energy than the national average of $1282? Use Excel to perform the hypothesis test.

SOLUTION We stored the sample data of Table 9.4 into a column named **ENERGY** in the **DATA** worksheet. Let μ denote the mean energy expenditure of all upper-income families in 1993 as contrasted to all families that year.

The problem is to perform the hypothesis test

H_0: $\mu = \$1282$ (mean was not greater than the national mean)

H_a: $\mu > \$1282$ (mean was greater than the national mean)

at the 5% significance level ($\alpha = 0.05$). Note that the hypothesis test is right-tailed since there is a greater-than sign (>) in the alternative hypothesis.

Since the standard deviation is unknown, we need to apply the *t*-test. An attempt to apply the *z*-test of the last section would fail at Step 2 since that information is nowhere to be found. Also, since the sample size is large, we do not need to check the data for normality. The **One Mean** panel in Figure 9.2 on the following page was produced by executing the Excel commands located below the figure. Observe that input cell **E3** has been cleared of any entry to prevent the *z*-test from being performed.

9.3 One-sample t-test for a population mean

FIGURE 9.2
Excel right-tailed test of ENERGY data

	One Mean		
DATA FILE NAME:		σ =	
ENERGY	n = 36	df = 35	
	xbar = 1455.917	Min = 932.00	
	s = 310.234	Q_1 = 1246.75	
	SE Mean 51.71	Median = 1407	
	$H_0: \mu_0$ = 1282.00	Q_2 = 1673.25	
	Alt Code = 1	Max = 2224.00	
t-Test	α = 0.05	t-Confidence	Interval:
Right-tailed test: t = 3.364		95%	One-sided:
Critical Value: 1.690		Lower Limit l =	1368.556
P-value: 0.0009		μ_0 =	1282.00
Action: Reject		**Action:**	Reject

Excel commands:

1. Type <u>ENERGY</u> in input cell **B4** under **DATA FILE NAME:** and press [Enter ↵]
2. Clear (delete) any entry in cell **E3** of a value for σ
3. Select input cell **D8**, type <u>1282</u> for the value of μ_0 and press [Enter ↵]
4. Type <u>1</u> for **Alt Code** in input cell **D9** and press [Enter ↵]
5. Type <u>.05</u> in input cell **D10** for the value of α and press [Enter ↵]

In the data summary, we find the sample size, sample mean, sample standard deviation, and estimated standard error of the mean. The results of interest to us here are displayed in the cell range **B10:F15**. Observe that the computed P-value of 0.0009 is less than the level of significance $\alpha = .05$. For a classical test, the test statistic has the value $t = 3.364$ and this is greater than the critical value of $t_{.05} = 1.690$. Finally, a 95% lower confidence bound for μ is 1368.556 and 1282 does NOT fall inside that interval, i.e. $1282 < 1368.556$.

In other words, the data provide sufficient evidence to conclude that upper-income families spent more, on the average, for energy in 1987 than the national average of \$1282. ■

Again, to perform this test for another set of data you have only to replace **ENERGY** with the name or the range (and a possible location) of the data and the results will be changed dynamically. Also, as previously mentioned, we often do not have the raw data with which to evaluate a study. In the open literature it is often the case that only summary statistics are presented. We may still perform an appropriate hypothesis test if we have only summary data as the next example shows.

EXAMPLE 9.3 Illustrates the t-test for a mean using summary data

As reported by the R.R. Bowker Company of New York in *Library Journal* the mean annual subscription rate to law periodicals was $97.33 in 1995. A sample of 12 law periodicals in 1996 yielded a sample average of $122.75 and a sample standard deviation of $8.44. At the 5% significance level, do the data indicate that the subscription rate has increased in 1996 over 1995? Use Excel to perform the hypothesis test.

SOLUTION The problem is to perform the hypothesis test

H_0: $\mu = \$97.33$ (mean was not greater in 1996)

H_a: $\mu > \$97.33$ (mean was greater in 1996)

at the 5% significance level ($\alpha = 0.05$). Note that the hypothesis test is right-tailed since there is a greater-than sign ($>$) in the alternative hypothesis.

Since the standard deviation is unknown, we need to apply the *t*-test. First, open the **DATA** worksheet and scroll to the upper left-hand corner. Column B has been given the name **SD** (for *summary data*). Proceed as follows.

Excel commands:

1. Type 12 in input cell **B2** next to **n ==>** for the sample size n
2. Select input cell **B3** and type 122.75 next to **xbar ==>** and press [Enter⏎]
3. Select input cell **B4** and type 8.44 next to **s ==>** and press [Enter⏎]
4. Click on the **1SAMP** tab and locate the panel **One Mean**
5. Type **SD** in input cell **B4** under **DATA FILE NAME:** and press [Enter⏎]
6. Clear input cell **E3** of any value for σ
7. Type 97.33 in input cell **D8** for μ_0
8. Type 1 for **Alt Code** in input cell **D9** and press [Enter⏎]
9. Type .05 in input cell **D10** for the value of α and press [Enter⏎]

The output (range **B10:F15**) is displayed in Figure 9.3.

FIGURE 9.3
Excel right-tailed test of summary data

t-Test		$\alpha =$	0.05	t-Confidence	Interval:
Right-tailed test: t =	10.433			95%	One-sided:
Critical Value:	1.796			Lower Limit l =	118.374
P-value:	0.0000			$\mu_0 =$	97.33
Action:	Reject			Action:	Reject

Observe that the computed *P*-value of 0.0000 is less than the level of significance $\alpha = .05$. For a classical test, the test statistic has the value $t = 10.433$ and this is considerably greater than the critical value of $t_{.05} = 1.796$. Finally, a 95% lower confidence bound for μ is 118.374 and 97.33 does NOT fall inside that interval, i.e., \$97.33 < \$118.37.

In other words, the data provide sufficient evidence (even moderately strong evidence) that the population mean subscription rate was greater in 1996 than in 1995. ∎

PROBLEM 9.2 As reported by the College Entrance Examination Board in *National College-Bound Senior*, the mean verbal score on the Scholastic Assessment Test (SAT) in 1995 was 428 points out of a possible 800. A random sample of 25 verbal scores for last year yielded the following data.

344	494	350	376	313
489	358	383	498	556
379	301	432	560	494
418	483	444	477	420
492	287	434	514	613

At the 10% significance level, does it appear that last year's mean for verbal SAT scores is greater than the 1995 mean of 428 points? The data are stored in a column named **SAT** in the **DATA** worksheet. ◇

LESSON 10

Inferences for Two Population Means

GENERAL OBJECTIVE In Lessons 8 and 9 we examined the Excel functions required to find confidence intervals and perform hypothesis tests for one population mean, μ. Frequently, however, inferential statistics is used to compare the means of *two* populations. This lesson shows how Excel can be employed to make inferences concerning two populations.

LESSON OUTLINE
10.1 Inferences for two population means (σs equal)
10.2 Inferences for two population means (σs not equal)
10.3 Inferences for two population means using paired samples

10.1 Inferences for two population means (σs equal)

E-547 : I-607

In the exercises in Section 10.1 of *IS/5e* and *ES/4e* (page 538 of *ES/4e* and page 598 of *IS/5e*) two-sample z-procedures are presented for inference concerning two means when the populations are normal and both population standard deviations, σ_1 and σ_2, are known. However, as discussed in your book, such circumstances are rare. It has always been standard practice when the sample sizes are very large, to apply these procedures substituting sample estimates of the two population standard deviations and treat the results as approximate.

When the population standard deviations are unknown, there are two methods that apply as long as the populations being sampled are *normally distributed*. One requires that the populations under consideration have equal standard deviations and the other does not. Our implementation of Excel can handle both of these cases.

Procedure 10.1 beginning on page 602 of *IS/5e* (page 542 of *ES/4e*) gives a step-by-step method for performing a hypothesis test to compare the means of two normally distributed populations with equal standard deviations or large sample sizes, using independent samples. Procedure 10.2 on page 607 of *IS/5e* (page 547 of *ES/4e*) determines confidence intervals for the difference of means in these same circumstances. Since it is assumed that the two standard deviations are equal, though unknown, the data from both samples are combined to estimate this common standard deviation. Hence, the procedures are called *pooled*. The next example illustrates the use of Excel for both of these situations.

EXAMPLE 10.1 *Illustrates the pooled t-test and pooled t-interval using Excel*

The annual salaries of faculty in public institutions is to be compared to the annual salaries of faculty in private institutions. A random sample of 30 salaries from public institutions and a random sample of 35 salaries from private institutions is taken. The salaries are displayed in Table 10.1.

TABLE 10.1 Annual salaries ($1000) for 30 faculty members in public institutions and 35 faculty members in private institutions

Public institutions						Private institutions					
34.2	63.6	24.4	79.4	33.8	88.2	92.9	102.2	51.5	77.6	71.1	59.3	71.0
90.0	56.8	56.0	42.4	40.2	44.6	52.0	102.2	51.5	77.6	71.1	59.3	71.0
100.4	41.4	58.2	81.8	51.2	64.4	63.1	53.8	45.2	78.3	67.6	27.2	92.6
24.6	35.0	76.8	29.2	41.2	74.0	118.5	101.0	76.0	66.3	52.4	81.2	56.0
107.4	54.2	84.2	15.8	60.2	71.0	37.7	68.6	56.1	31.3	47.2	24.8	62.3

At the 5% significance level, does it appear that there is a difference in mean salaries for faculty teaching in public and private institutions? Use Excel to make the decision and assume that the standard deviations are equal.

SOLUTION Let μ_1 and μ_2 denote the actual mean salary for faculty in public and private institutions, respectively. Then the problem is to perform the hypothesis test

H_0: $\mu_1 = \mu_2$ (mean salaries are the same)

H_a: $\mu_1 \neq \mu_2$ (mean salaries are different)

at the 5% level of significance. Note that the test is two-tailed since there is a not-equal sign ($\neq$) in the alternative hypothesis.

By assumption, the samples are independent and the sample sizes are both large. Hence we are entitled to use Procedure 10.1. We have entered the two sets of sample data from Table 10.1 into columns named **PUBLIC** and **PRIVATE** in the **DATA** worksheet of the Inference workbook. So, click on the **2SAMP** worksheet tab, locate the **Two Means** panel, and proceed as follows.

Excel commands:

1 Under the heading **DATA FILE NAMES:**, type PUBLIC in input cell **B4**, press the ↓ key to input cell **B5**, type PRIVATE and press Enter

2 Select input cell **D7**, type 0, select input cell **F7** and type y (or Y) to indicate the standard deviations are equal

3 Select input cell **D8** and type 0 to indicate a two-tailed test

4 Select input cell **F8**, type .05 and press Enter

The results are displayed in the cell range **B9:F13** instantly and are shown in Figure 10.1 along with other items of interest.

FIGURE 10.1 Excel pooled t-test for equality of means

	Two Means			
DATA FILE NAMES:	n_1 =	30	n_2 =	35
PUBLIC	$\bar{x}_1$ =	57.480	$\bar{x}_2$ =	66.394
PRIVATE	s_1 =	23.953	s_2 =	22.261
	s_p =	23.055	df =	63
	H_0: μ_d =	0.00	σ's Equal ?	y
Pooled	Alt Code	0	α =	0.05
Two-tailed t-test: t=	-1.554		95%	**Two-sided:**
Lower Critical Value:	-1.998		Lower Limit l =	-20.377
Upper Critical Value:	1.998		Upper Limit u =	2.549
P-value:	0.1252		μ_d =	0
Action:	Do not Reject		Action:	Do not Reject

Observe that the computed P-value of 0.1252 is greater than the level of significance $\alpha = .05$. For a classical test, the test statistic has the value $t = -1.554$ and this is between the critical values, that is, $-1.998 < -1.554 < 1.998$. Finally, a 95% confidence interval runs from -20.377 to $+2.549$ and that includes 0. You are informed that, by all these criteria, you should not reject the null hypothesis. In other words, the data do not provide sufficient evidence that there is a difference between the mean salaries of faculty teaching in public and private institutions. ∎

Incidentally, the theory allows us to test the hypothesis that the mean difference is any constant, not just 0. This will be true of all of the procedures we discuss in this lesson. While a difference of 0 is by far the most common use of the procedures, we can imagine scenarios when we might already acknowledge a difference of means and would like to make a decision on how much that difference is. All you need to do in that case is replace 0 by the constant of interest in input cell **D7** at Step 2.

PROBLEM 10.1 In a packing plant, a machine packs cartons with jars. A salesperson claims that the machine she is selling will pack faster. To test that claim, the time it takes each machine to pack 10 cartons is recorded. The results, in seconds, are in the table that follows.

New machine		Present machine	
42.0	41.0	42.7	43.6
41.3	41.8	43.8	43.3
42.4	42.8	42.5	43.5
43.2	42.3	43.1	41.7
41.8	42.7	44.0	44.1

Do the data suggest that the new machine packs *faster*, on the average? Use $\alpha = 0.05$. [Assume the packing times for both machines are normally distributed and that the standard deviations of the packing times for both machines are the same.] Employ Excel to perform the appropriate hypothesis test. The data are stored in the **DATA** worksheet under the names **NEW** and **PRESENT**. ⋄

10.2 Inferences for two population means (σs not equal)

E-560 : I-619

Section 10.3 starting on page 615 of *IS/5e* (page 554 of *ES/4e*) discusses inference to compare the means of two normally distributed populations using independent samples without making any assumptions about the relationship between the standard deviations of the two populations. The basis for the inference is the following key fact.

KEY FACT 10.1

Suppose that independent random samples of sizes n_1 and n_2 are to be taken from two normally distributed populations with means μ_1 and μ_2, respectively.

Then the random variable

$$T = \frac{(\overline{X}_1 - \overline{X}_2) - (\mu_1 - \mu_2)}{\sqrt{(S_1^2/n_1) + (S_2^2/n_2)}}$$

has approximately the t-distribution with degrees of freedom given by

$$\Delta = \frac{\left[(s_1^2/n_1) + (s_2^2/n_2)\right]^2}{\dfrac{(s_1^2/n_1)^2}{n_1 - 1} + \dfrac{(s_2^2/n_2)^2}{n_2 - 1}}$$

rounded down to the nearest integer.

Thus the test statistic for a hypothesis test with null hypothesis $H_0 : \mu_1 = \mu_2$, is

$$t = \frac{(\overline{x}_1 - \overline{x}_2)}{\sqrt{(s_1^2/n_1) + (s_2^2/n_2)}},$$

and the endpoints of a $(1 - \alpha)$-level confidence interval for $\mu_1 - \mu_2$ are

$$(\overline{x}_1 - \overline{x}_2) \pm t_{\alpha/2} \cdot \sqrt{(s_1^2/n_1) + (s_2^2/n_2)}.$$

We will demonstrate this application in the next example using Excel. Actually, the steps to perform the nonpooled test have been arranged in the **2SAMP** worksheet so that the only change in the input is to type **n** (or anything other than **y**) in cell **F7** at Step 2 of the Excel instructions given for Example 10.1. In order to vary the input procedure slightly, let us apply a summary data option. For this we will use the summary data provided for Example 10.6 on page 616 of *IS/5e* (page 557 of *ES/4e*).

EXAMPLE 10.2 *Illustrates the nonpooled two-sample t-procedure using summary data*

A group of neurosurgeons wanted to see whether a dynamic system (Z-plate) reduced the operative time relative to a static system (ALPS plate). R. Jacobowitz, Ph.D., an ASU professor, along with G. Vishteh, M.D., and other neurosurgeons, obtained the following data on operative times, in minutes.

TABLE 10.2
Summary statistics for operative times for dynamic and static systems

Dynamic	Static
$n_1 = 14$	$n_2 = 6$
$\overline{x}_1 = 394.6$	$\overline{x}_2 = 468.3$
$s_1 = 84.7$	$s_2 = 38.2$

At the 1% level of significance, do the data provide sufficient evidence to conclude that the mean operative time is less with the dynamic system than with the static system?

10.2 Inferences for two population means (σ's not equal)

SOLUTION Open the **DATA** worksheet in the Inference notebook and scroll to the beginning of the sheet for summary data entry into the two columns labeled SD and SD2. Enter the two sets of data as follows.

Excel commands:

1 Type <u>14</u> in input cell **B2** next to **n** ==> for sample size n_1, select input cell **D2** and type <u>6</u> for n_2

2 Select input cell **B3** and type <u>394.6</u> next to **xbar** ==> for $\bar{x}_1$, select input cell **D3** and type <u>468.3</u> for $\bar{x}_2$

3 Select input cell **B4** and type <u>84.7</u> next to **s** ==> for s_1, select input cell **D4**, type <u>38.2</u> for s_2 and press [Enter⏎]

To carry out the test, open up the **2SAMP** worksheet by clicking on its tab, locate the **Two Means** panel and proceed as follows

Excel commands:

1 Type <u>SD</u> in input cell **B4** under **DATA FILE NAMES:**, select input cell **B5** and type <u>SD2</u>

2 Type <u>0</u> if necessary in input cell **D7** and type <u>n</u> in input cell **F7**

3 Type <u>-1</u> for **Alt Code** in input cell **D8**, type <u>.01</u> in input cell **F8** for the value of α and press [Enter⏎]

The results are displayed in the cell range B9:F13 as shown in Figure 10.2.

FIGURE 10.2
Excel nonpooled
t-test for
equality of means

		Two Means		
DATA FILE NAMES:	n_1 = 14		n_2 = 6	
SD	$xbar_1$ = 394.600		$xbar_2$ = 468.300	
SD2	s_1 = 84.700		s_2 = 38.200	
	s = 27.489		df = 17	
	H_0: μ_d = 0.00		σ's Equal ?	n
Nonpooled	Alt Code: -1		α =	0.01
Left-tailed t-test: t =	-2.681		**99%**	**One-sided:**
Critical Value:	-2.567			
			Upper Limit u =	-3.138
P-value:	0.0079		μ_d =	0
Action:	Reject		**Action:**	Reject

Observe that the computed *P*-value of 0.0079 is less than the level of significance $\alpha = 0.01$. For a classical test, the test statistic has the value $t = -2.681$ and this is less than the critical value of $-t_{.01} = -2.567$. Finally, a 99% upper confidence bound for μ is -3.138 and 0 does NOT fall inside that interval, i.e., $0 > -3.138$. You are informed that by all these criteria that you should reject the null hypothesis. In other words, the data provide sufficient evidence that the dynamic system reduces the mean operative time. ∎

PROBLEM 10.2 The owner of a chain of car washes needs to decide between two brands of hot waxes. One of the brands, Sureglow, costs less than the other brand, Mirror-Sheen. Therefore, unless there is strong evidence that the second brand outlasts the first, the owner will purchase the first brand. With the cooperation of several local automobile dealers, 30 cars are randomly selected to take part in the test. Fifteen of the 30 cars are waxed with Sureglow and 15 with Mirror-Sheen. The cars are then exposed to the same environmental conditions. The test results below show the give the effectiveness times, in days.

Sureglow			Mirror-Sheen		
87	90	88	92	92	91
93	90	92	91	92	91
91	89	93	93	93	92
88	87	89	92	93	94
91	91	90	93	94	91

At the 1% significance level, does Mirror-Sheen seem to have a longer effectiveness time, on the average, than Sureglow? [Assume that the effectiveness times for both waxes are normally distributed.] The data are stored in columns **SURE** and **MIRROR** in the **DATA** worksheet. ◇

10.3 Inferences for two population means using paired samples

Up to now we have considered inferences for two population means when the samples are *independent*. However, it is often more suitable to use *paired samples*. With paired samples, each item in the sample consists of a pair of numbers, one from each of the two populations. We can think of the paired differences of the pairs sampled as a random sample from the population of all possible paired differences. If that population is *normally distributed*, then we can apply the following key fact for deriving tests and confidence intervals.

KEY FACT 10.2

Suppose that a random sample of n pairs is to be taken from populations with means μ_1 and μ_2. Further suppose that the population of all paired differences, represented by the random variable D, is normally distributed. Then the random variable

$$T = \frac{\overline{D} - (\mu_1 - \mu_2)}{S_d/\sqrt{n}}$$

has the *t*-distribution with df $= n - 1$.

INFERENCE FOR TWO POPULATION MEANS USING PAIRED SAMPLES

E-577 : I-649

Procedure 10.6 beginning on page 645 of *IS/5e* (Procedure 10.5 on page 572 of *ES/4e*) gives a step-by-step method for performing a hypothesis test to compare the means of two normally distributed populations when the samples are paired. Procedure 10.7 on page 648 of *IS/5e* (Procedure 10.6 on page 576 of *ES/4e*) determines confidence intervals for the difference of means in these same circumstances. With Excel it is a simple matter to carry out both of these procedures since the problem reduces to a one sample problem, once sample differences have been gathered.

EXAMPLE 10.3 *Illustrates the use of paired samples to perform a paired-difference test*

A major oil company has developed a new gasoline additive that is supposed to increase mileage. To test that hypothesis, 10 cars are randomly selected. Each car sampled is driven both with and without the additive. The resulting gas mileages, in miles per gallon, are displayed in the second and third columns of Table 10.3 that we present next.

TABLE 10.3 Gas mileages, with and without additive, for 10 randomly selected cars

Car	With additive x_1	Without additive x_2
1	25.7	24.9
2	20.0	18.8
3	28.4	27.7
4	13.7	13.0
5	18.8	17.8
6	12.5	11.3
7	28.4	27.8
8	8.1	8.2
9	23.1	23.1
10	10.4	9.9

Do the data suggest that, on the average, the gasoline additive improves mileage? Use Excel to perform the appropriate hypothesis test at the 5% level of significance. [Assume that changes in gas mileage due to the gasoline additive are normally distributed.]

SOLUTION Let μ_1 denote the mean gas mileage of all cars when the additive is used and μ_2 denote the mean gas mileage of all cars when the additive is not used. Then we want to perform the hypothesis test

$H_0: \mu_1 = \mu_2$ (mean mileage with additive is not greater)

$H_a: \mu_1 > \mu_2$ (mean mileage with additive is greater)

at the 5% significance level. Note that the samples are *paired*. Each item in the sample consists of a pair of numbers; namely, the gas mileage of a given car both with and without the additive.

We entered the paired samples into columns named **ADD** and **NOADD** in the **DATA** worksheet of the Inference workbook. Then we simply created a new column named **DIFFER** of differences ADD-NOADDD. With that much done, we merely execute the t-test of the last chapter with the standard deviation unknown of course. Open the **1SAMP** worksheet. Here are the steps and output.

Excel commands:

1 Type <u>DIFFER</u> in input cell **B4** under **DATA FILE NAME:** and press [Enter⏎]

2 Clear (delete) any entry in input cell **E3** of a value for σ

3 Select input cell **D8** and type <u>0</u> for the value of μ_0 and press [Enter⏎]

4 Type <u>1</u> for **Alt Code** in input cell **D9**, type <u>.05</u> in input cell **D10** for the value of α and press [Enter⏎]

The results are displayed in the cell range **B10:F15** and are reproduced in Figure 10.3.

FIGURE 10.3
Excel right-tailed test for paired differences

t-Test		α = 0.05	t-Confidence	Interval:
Right-tailed test: t =	4.714		**95%**	**One-sided:**
Critical Value:	1.833		**Lower Limit l =**	0.403
P-value:	0.0005		μ_0 =	0.00
Action:	Reject		**Action:**	Reject

Observe that the computed P-value of 0.0005 is less than the level of significance $\alpha = .05$. For a classical test, the test statistic has the value $t = 4.714$ and this is greater than the critical value of $t_{.05} = 1.833$. Finally, a 95% lower confidence bound for μ is 0.403 and 0 does NOT fall inside that interval.

Thus, the data provide sufficient evidence to conclude that the mean gas mileage of all cars when the additive is used is greater than the mean gas mileage of all cars when the additive is not used. In other words, it appears that the additive is effective in increasing gas mileage. ∎

PROBLEM 10.3 An agronomist has developed a new variety of wheat that she feels will return a greater yield. For an initial test, she selects 10 wheat farms at random. On each farm, two adjacent one-acre plots are chosen. A widely-used current variety is planted on one of the one-acre plots and the new variety on the other. The yields, in bushels, are shown in the following table.

Farm	Current variety x_1	New variety x_2
1	34	37
2	40	35
3	34	34
4	33	34
5	42	44
6	35	39
7	34	33
8	42	44
9	41	43
10	39	40

Assume that changes in crop yields due to wheat varieties are normally distributed.

Do the data provide evidence, at the 5% significance level, that the new variety of wheat gives a better yield, on the average, than the current variety? Use Excel to find the answer. We stored the data in columns named **CURRENT** and **NEWV**. You may use the column **DIFFER** again, or use a different name. ◇

LESSON 11

Inferences for Population Standard Deviations

GENERAL OBJECTIVE In previous lessons, we have examined the Excel functions that may be used for statistical inferences concerning population means. Now we will learn how Excel can be applied to perform tests and compute confidence intervals for standard deviations.

LESSON OUTLINE
11.1 The χ^2-distribution
11.2 Inferences for one population standard deviation
11.3 Inferences for two population standard deviations

11.1 The χ^2 distribution

I-687

The inferential procedures discussed in this lesson rely on a class of continuous probability distributions called *chi-square distributions*. Probabilities for a random variable having a chi-square distribution are equal to areas under a curve—a χ^2 (chi-square) curve.

To perform a hypothesis test or obtain a confidence interval that is based on a chi-square distribution, we will need to be able to find the χ^2-value(s) corresponding to a specified area under a χ^2-curve. Recall that the symbol χ^2_α is used to represent the χ^2-value with area α to its right under a χ^2-curve. This quantity can be found using Table V in *IS/5e* or, alternatively, with the aid of Excel. In addition, we need to compute the area under a χ^2 curve in order to compute P-values for test statistics.

Up to now the distributions we have been dealing with for statistical inference have been symmetric. That is, the area to the right of the mean is a mirror image of the area to the left. This is not true with the χ^2 family of distributions (see Figure 11.1 below). Like the t family, each χ^2 distribution is distinguished by a parameter known as degrees of freedom, a positive whole number that we will have to supply to the Excel functions in order to produce a desired output.

FINDING THE χ^2-VALUE FOR A SPECIFIED AREA

Excel has a function called CHIINV which, with a specification of the degrees of freedom, may be used to compute the χ^2-value with a specified area to its *right*. From this, it is a simple matter to obtain the χ^2-value(s) corresponding to any specified area under an arbitrary χ^2-curve. We present the details in the example to follow.

EXAMPLE 11.1 *Illustrates how to find the χ^2-value for a specified area using Excel*

For a χ^2-curve with 20 degrees of freedom, find $\chi^2_{0.05}$; that is, find the χ^2-value with area 0.05 to its right. See Figure 11.1

FIGURE 11.1
The χ^2-curve with 20 degrees of freedom

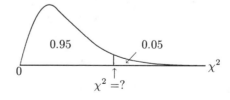

118 Lesson 11 Inferences for Population Standard Deviations

SOLUTION As we mentioned, the function CHIINV, with a specification of degrees of freedom, computes the χ^2-value with a specified area to its *right*. So, to find the χ^2 value with area 0.05 to its right, simply select any empty cell in any Excel worksheet and type =CHIINV(.05,20) as a formula in that cell. You should find the value 31.4104 in the cell instantly. That is, the output shows that $\chi^2_{0.05} = 31.41042$. ∎

PROBLEM 11.1 For a χ^2-curve with 14 degrees of freedom, use Excel to find
a) the χ^2-value with area 0.10 to its right.
b) the χ^2-value with area 0.10 to its left. ◊

FINDING THE AREA FOR A GIVEN χ^2-VALUE

Using the CHIDIST function in Excel, we can find the area to the right of any specified value of χ^2. The format is CHIDIST(x,df) and the output is the area to the *right* of x under a χ^2-curve having degrees of freedom df. Hence we can find other areas as well.

EXAMPLE 11.2 *Illustrates how to find the area under a χ^2-curve for a specified value using Excel*

For a χ^2-curve with 20 degrees of freedom, find the area to the right of the value $\chi^2 = 23$. See Figure 11.2.

FIGURE 11.2
The χ^2-curve with 20 degrees of freedom

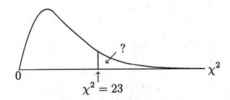

SOLUTION Select any empty cell in Excel and type =CHIDIST(23.0,20). The cell will display the value 0.2888 instantly. Hence the area to the right of $\chi^2 = 23$ for this particular χ^2-curve is approximately 0.29. ∎

PROBLEM 11.2 For a χ^2-curve with 14 degrees of freedom, use Excel to find
a) the area to the right of $\chi^2 = 13$.
b) the area between $\chi^2 = 7$ and $\chi^2 = 13$. ◊

11.2 Inferences for one population standard deviation

Recall that the *standard deviation*, σ, of a population is a measure of dispersion or variability of the population values. Populations with large standard deviations have a large spread, while those with small standard deviations have a small spread.

There are many circumstances where the standard deviation of a population is unknown, but where it is nevertheless important to have information about its value. In most cases, it is impossible to determine the value of the population standard deviation exactly. Consequently, we employ inferential statistics to obtain the required information.

In this section, we will learn how Excel can be used to help carry out the inferential procedures for a population standard deviation that are presented in Section 11.1 of *IS/5e*. We begin with hypothesis tests.

HYPOTHESIS TESTS FOR A POPULATION STANDARD DEVIATION

I-693

Excel does not have a specific function to perform a hypothesis test for a population standard deviation. However, we can use Excel primitive functions to do the calculations required in implementing Procedure 11.1 beginning on page 691 of *IS/5e*. Consider the following example where the steps are repeated.

EXAMPLE 11.3 *Illustrates how Excel can be used in conjunction with Procedure 11.1*

A hardware manufacturer produces 10-millimeter bolts. The manufacturer needs to decide whether the standard deviation, σ, of bolt diameters is *less than* 0.09 mm. He randomly samples 12 bolts and obtains their diameters. The results are shown in Table 11.1.

TABLE 11.1
Diameters, in mm, of 12 randomly selected bolts

10.05	10.00	10.02	9.97
10.07	10.03	9.98	10.10
9.95	9.99	10.00	10.08

At the 5% significance level, do the data provide evidence that the standard deviation, σ, of bolt diameters is less than 0.09 mm? [Assume that the population of diameters of all bolts manufactured is normally distributed.] Use Excel to compute the results.

SOLUTION We have stored the data in a column labeled **DIAMETERS** in the **DATA** worksheet of the Inference workbook. We first apply Procedure 11.1 step-by-step as in your book, using Excel to assist in the calculations required.

STEP 1 *State the null and alternative hypotheses.*

The null and alternative hypotheses are

$$H_0: \sigma = 0.09 \text{ mm (too much variation)}$$
$$H_a: \sigma < 0.09 \text{ mm (not too much variation)}$$

Note that the test is left-tailed since there is a less-than sign (<) in the alternative hypothesis.

STEP 2 *Decide on the significance level, α.*

The test is to be performed at the 5% significance level. Thus, $\alpha = 0.05$.

STEP 3 *The critical value for a left-tailed test is $\chi^2_{1-\alpha}$, with $df = n - 1$.*

We have $\alpha = 0.05$ and $n = 12$. Consequently, $df = 12 - 1 = 11$. Therefore, we need to determine $\chi^2_{1-0.05} = \chi^2_{0.95}$ for $df = 11$. Since $\chi^2_{0.95}$ is the χ^2-value with area 0.95 to its right this χ^2-value can be obtained by applying the function CHIINV, as explained previously. Select any empty cell in an Excel worksheet and type =CHIINV(.95,11) as a formula in that cell. You should find the value 4.5748 in the cell instantly.

STEP 4 *Compute the value of the test statistic*

$$\chi^2 = \frac{n-1}{\sigma_0^2} s^2$$

First we have Excel compute the sample standard deviation, s, of the data in Table 11.1. In any empty cell of any Inference worksheet, type the following command: =(12-1)*(STDEV(DIAMETERS))^2/(0.09)^2. The result 2.98765 appears immediately.

STEP 5 *If the value of the test statistic falls in the rejection region, reject H_0; otherwise, do not reject H_0.*

From Step 4, the value of the test statistic is $\chi^2 = 2.98765$. Since this value lies to the left of the critical value $\chi^2_{0.95} = 4.5748$ and the test is left-tailed, we reject H_0. Alternatively, we can find the P-value of the test using Excel. Just type 1-CHIDIST(2.98765,11) in any empty cell and the value 0.0091 appears in that cell. Since the test is left-tailed, this is the P-value of the test. Accordingly, since we have $0.0091 < 0.05$, H_0 should be rejected at the 5% level of significance.

STEP 6 *State the conclusion in words.*

It appears that the standard deviation, σ, of the diameters of all bolts being produced is less than 0.09 mm. ∎

With all of the steps involved in Procedure 11.1, this is a natural place to store them in a worksheet for repeated dynamic results. We have done just that in a panel labeled **One Standard Deviation** in the **1SAMP** worksheet. To implement this

11.2 Inferences for one population standard deviation

panel, open that worksheet, locate the panel, and proceed much like the previous chapters as follows.

Excel commands:

1. Type **DIAMETERS** under **DATA FILE NAME:** in the input cell **Q4**
2. Select input cell **S6** and type **.09**
3. Select input cell **S7** and type **-1** for the **Alt Code**
4. Select input cell **S8**, type **.05** and press ⏎

Figure 11.3 shows the resulting output located in the range **Q9:R13**.

FIGURE 11.3
Excel left-tailed test of DIAMETERS data

χ^2-Test Stat :	2.988
Critical Value:	4.575
P-value:	0.0091
Action:	Reject

This output confirms the tedious calculations performed for this test. ∎

PROBLEM 11.3 Every year thousands of college-bound high-school students take the *Scholastic Aptitude Test* (SAT). This test, provided by the Educational Testing Service of Princeton, New Jersey, measures the verbal and mathematical abilities of prospective college students. Student scores are reported on a scale that ranges from a low of 200 to a high of 800. This scale was introduced in 1941. At that time, the standard deviation of scores was 100 points. Suppose that a random sample of 25 verbal scores for this year gives the following results:

560	405	367	416	540
347	370	629	570	372
396	439	526	610	339
569	314	228	483	353
379	558	432	648	475

At the 5% significance level, do the data suggest that the standard deviation, σ, for this year's verbal scores is *different* from the 1941 standard deviation of 100? The data are stored in a column named **VERBAL** in the **DATA** worksheet. ◊

CONFIDENCE INTERVALS FOR A POPULATION STANDARD DEVIATION   I-695

We can also use Excel to aid us in constructing a confidence interval for a population standard deviation, σ. In the next example, we will see how Excel can be applied to help perform Procedure 11.2 on page 694 of *IS/5e*.

EXAMPLE 11.4 *Illustrates how Excel can be used in conjunction with Procedure 11.2*

Refer to Example 11.3 on page 119. Use the sample data in Table 11.1 on that page to find a 95% confidence interval for the standard deviation, σ, of the diameters of all bolts produced by the manufacturer. Employ Excel in conjunction with Procedure 11.2 of *IS/5e*.

SOLUTION We repeat the steps of Procedure 11.2 here.

STEP 1 *For a confidence level of $1 - \alpha$, find $\chi^2_{1-\alpha/2}$ and $\chi^2_{\alpha/2}$ for $df = n - 1$.*

We want a 95% confidence interval. This means $\alpha = 0.05$. Also, $n = 12$, which makes $df = 12 - 1 = 11$. Now,

$$\chi^2_{\alpha/2} = \chi^2_{0.05/2} = \chi^2_{0.025}$$

and

$$\chi^2_{1-\alpha/2} = \chi^2_{1-0.05/2} = \chi^2_{0.975}$$

We will use Excel to find these two χ^2-values. Since $\chi^2_{0.975}$ is the χ^2-value with area 0.975 to its right, this number can be obtained by applying the function CHIINV, as explained previously. Select any empty cell in an Excel worksheet and type =CHIINV(.975,11) as a formula in that cell. You should find the value 3.8157 in the cell instantly. Similarly, $\chi^2_{0.025}$ is the χ^2-value with area 0.025 to its right. This number can be obtained by selecting any empty cell in an Excel worksheet and typing =CHIINV(.025,11) as a formula in that cell. You should find the value 21.92 in the cell.

STEP 2 *The confidence interval for σ is*

$$\sqrt{\frac{n-1}{\chi^2_{\alpha/2}}} \cdot s \quad to \quad \sqrt{\frac{n-1}{\chi^2_{1-\alpha/2}}} \cdot s$$

In Example 11.3, we found the sample standard deviation, s, of the bolt diameters to be 0.0469. In any two neighboring empty cells, type

LET C12=SQRT((12-1)/21.92)*.0469

and

LET C12=SQRT((12-1)/21.92)*.0469.

The results in the cells are the 95% confidence limits and run from 0.03322 to 0.07963. We can be 95% confident that the standard deviation, σ, of all bolt diameters is somewhere between 0.0332267 mm and 0.0796377 mm. ∎

You should have noticed in the last example that the **One Standard Deviation** panel automatically computed a confidence interval, one that was compatible with the one-sided test in the example. The entire panel is shown in Figure 11.4 on the following page.

FIGURE 11.4
Excel output of
DIAMETERS data

	One Standard Deviation			
DATA FILE NAME:	n =	12	Min =	9.95
DIAMETERS	s =	0.0469	Q_1 =	9.9875
	df =	11	Median =	10.01
	H_0: σ_0 =	0.09	Q_2 =	10.055
	Alt Code =	-1	Max =	10.10
	α =	0.05		
χ^2-Test Stat :	2.988		χ^2-Confidence	Interval:
Critical Value:	4.575			
			Upper Limit u =	0.073
P-value:	0.0091		σ_0 =	0.09
Action:	Reject		Action:	Reject

Since the test was left-tailed, we see that the upper one-sided 95% confidence bound on σ is 0.073. Not surprisingly, we reject the null hypothesis since the hypothesized standard deviation falls outside the confidence interval, that is, .09 > .073. ∎

PROBLEM 11.4 Refer to Problem 11.3 on page 121. Determine a 95% confidence interval for the standard deviation, σ, of this year's verbal SAT scores. Use Excel in conjunction with Procedure 11.2 of *IS/5e*. ◊

11.3 Inferences for two population standard deviations

Tests and confidence intervals for one mean extended to the same considerations for two means. Hence one would expect an extension of the last section to the case of two standard deviations. In the case of two means, the theoretical structure that yielded the tests was based on sums or differences of random variables. This led to tests and confidence intervals for the difference of two means in Lesson 10.

In the case of standard deviations, it is the ratio of two random variables that yields the structure necessary for tests and confidence intervals. Hence we focus on the ratio of two populations standard deviations, rather than the difference. We get the same information concluding that the ratio is 1 as we do from concluding that a difference is 0. It does necessitate exploring still another probability distribution called the F family.

THE F-DISTRIBUTION [I-703]

To perform a hypothesis test or obtain a confidence interval that is based on an F distribution, we will need to be able to find the F-value(s) corresponding to a specified area under an F-curve. Recall that the symbol F_α is used to represent the F-value with area α to its right under an F-curve. This quantity can be found with

124 Lesson 11 Inferences for Population Standard Deviations

the aid of Excel. In addition, we need to compute the area under an F curve in order to compute P-values for test statistics. We explain the details in the following examples.

EXAMPLE 11.5 Illustrates finding F percentiles

For an F-curve with df $= (4, 12)$, find $F_{0.05}$. That is, find the F-value with area 0.05 to its right.

SOLUTION The function FINV computes the F-value with a specified area to its *right*. The format is FINV(α, df_1, df_2). We want to find the F-value having area 0.05 to its *right* for an F-curve with df $= (4, 12)$. Find any empty cell and type `FINV(.05,4,12)` and discover that, for an F-curve with df=(4,12), $F_{0.05} = 3.2592$ ∎

EXAMPLE 11.6 Illustrates finding the area for specified F-values

For an F-curve with df $= (4, 12)$, find the area to the right of $F = 1.8$.

SOLUTION The function FDIST computes the area to the *right* of a given F-value. The format is FDIST(x, df_1, df_2). We want to find the area to the *right* of $F = 1.8$ with df=(4,12). Find any empty cell and type `FDIST(1.8,4,12)` and discover that, for an F-curve with df=(4,12), the area to the right of $F=1.8$ is 0.1937. ∎

PROBLEM 11.5

For an F-curve with df $= (12, 5)$, find
a) $F_{0.05}$ b) $F_{0.01}$ c) $F_{0.025}$
d) Find the area to the right of $F = 1.5$ under this curve   ◇

THE F-TEST FOR TWO POPULATION STANDARD DEVIATIONS I-709

Procedure 11.3 on page 708 of *IS/5e* carries out the steps for an F-test of the hypothesis that two standard deviations are equal. Excel has no set of functions for carrying out this procedure directly. However, we have incorporated several Excel functions in the **2SAMP** worksheet of the Inference workbook to accomplish the task. We illustrate with Example 11.14 of *IS/5e*.

EXAMPLE 11.7 Illustrates Procedure 11.3 for an F-test

In the 1997 paper "Using Repeatability and Reproducibility Studies to Evaluate a Destructive Test Method" (*Quality Engineering*, 10(2), pp.283–290), A. Phillips et al., studied the variability of the Elmendorf tear test. That test is used to evaluate material strength for fiberglass shingles, paper quality, and other manufactured products.

11.3 Inferences for two population standard deviations

In one aspect of the study, the researchers independently and randomly obtained data on Elmendorf tear strength of three different vinyl floor coverings. Table 11.2 provides the data, in grams, for two of the three vinyl floor coverings.

TABLE 11.2 Results of Elmendorf tear test on two different vinyl floor coverings

Brand A		Brand B	
2288	2384	2592	2384
2368	2304	2512	2432
2528	2240	2576	2112
2144	2208	2176	2288
2160	2112	2304	2752

Apply Procedure 11.3 to test the hypothesis that the two population standard deviations are equal at the 5% level of significance using Excel.

SOLUTION We entered the data into the **DATA** worksheet in two columns named BRANDA and BRANDB. We need to carry out the following test.

H_0: $\sigma_1 = \sigma_2$ (standard deviation of tear strengths are the same)

H_a: $\sigma_1 \neq \sigma_2$ (standard deviation of tear strengths are different)

Note that the test is two-tailed since there is a not-equal sign ($\neq$) in the alternative hypothesis. Open the **2SAMP** worksheet in the Inference workbook and locate the **Two Standard Deviations** panel. We obtained the output displayed in Figure 11.5 by executing the Excel commands below the figure.

FIGURE 11.5 Excel output of 2SAMP panel for BRANDA and BRANDB data

	Two Standard Deviations			
DATA FILE NAMES:	n_1 =	10	n_2 =	10
BRANDA	s_1^2 =	16466.489	s_2^2 =	39867.733
BRANDB	s_1 =	128.322	s_2 =	199.669
	df_1 =	9	df_2 =	9
	$H_0: \sigma_1/\sigma_2$ =	1.00	s_1/s_2 =	0.6427
	α =	0.05	Alt Code =	0
F-Test Stat :	0.413		F-Conf. Interval	for σ_1/σ_2
Lower Critical Value:	0.248		Lower Limit l =	0.320
Upper Critical Value:	4.026		Upper Limit u =	1.290
P-value:	0.2039		σ_1/σ_2 =	1.00
Action:	Do not Reject		Action:	Do not Reject

Excel commands:

1 Under **DATA FILE NAMES**: type BRANDA in input cell **O4** and type BRANDB in input cell **O5**

2 Type 1 in input cell **Q7** and type .05 in input cell **Q8**

3 Type 0 in input cell **S8** for Alt Code and press ⏎

The test results are not statistically significant at the 5% level of significance. For one thing, the P-value of 0.2039 is larger than $\alpha = 0.05$. The value of the test statistic, 0.413, lies between the lower and upper critical values and thus is not in the rejection region. Finally, since the confidence interval for the ratio (see next section) is already computed, notice that the hypothesized value of 1 lies inside the confidence interval.

Thus, the data do not provide sufficient evidence to conclude that the population standard deviations of tear strength differ for the two floor coverings. ∎

PROBLEM 11.6 Does cloud seeding makes a difference? In a study on the effects of cloud seeding in Tasmania, data were collected for various sections of the country at different seasons of the year. A partial list of reported rainfalls in inches for Eastern Tasmania during the summer season is summarized in the following table (S=seeded, U=unseeded).

S		U	
0.40	0.86	0.78	2.16
1.79	0.00	0.73	0.19
1.13	0.17	0.88	0.25
0.44	0.96	0.31	1.04
0.66	0.46	1.59	0.68
0.22	1.76	1.11	5.12

Is there statistical evidence to show that rainfall is more evenly spread out (less variable) in regions that were seeded over those that were not? The data have been saved in the **DATA** worksheet under the names **S** and **U**. Use the 5% level of significance.

CONFIDENCE INTERVALS FOR THE RATIO OF TWO POPULATION STANDARD DEVIATIONS $\boxed{\text{I-711}}$

We can also use Excel to aid us in constructing a confidence interval for two population standard deviations following the steps of Procedure 11.4 on page 710 of *IS/5e*. Indeed, we have already incorporated the elements of this procedure in the **Two Standard Deviations** panel to find a confidence interval for the ratio σ_1/σ_2.

EXAMPLE 11.8 *Illustrates how Excel can be used in conjunction with Procedure 11.4*

Refer to Example 11.7 on page 124. To vary the problem slightly and illustrate using summary data, suppose that you only have the information $n_1 = 10 = n_2$, $s_1 = 128.32$ and $s_2 = 199.67$. Employ the **Two Standard Deviations** panel to find a 95% confidence interval for the ratio σ_1/σ_2.

11.3 Inferences for two population standard deviations

SOLUTION Open the **DATA** worksheet of the Inference worksheet and proceed as follows to enter the data as summary data in columns B and D.

Excel commands:

1. Select input cell **B2** and type <u>10</u>
2. Select input cell **B4** and type <u>128.32</u>
3. Select input cell **D2** and type <u>10</u>
4. Select input cell **D4**, type <u>199.67</u> and press [Enter⏎]

Now open the **2SAMP** worksheet by clicking on its tab. Complete the input as follows.

Excel commands:

1. Select input cell **O4** and type <u>SD</u>
2. Select input cell **O5**, type <u>SD2</u> and press [Enter⏎]

All other inputs being as they were in the previous example, you will see that the output is precisely the same as that shown in Figure 11.5, except for the DATA FILE NAMES:. Among other things, this tells us not to reject the hypothesis of equality ($\sigma_1/\sigma_2 = 1$) because the value 1 is included in the confidence interval. ■

PROBLEM 11.7 Refer to Problem 11.6 on page 126. Determine a 95% upper confidence bound for the ratio of standard deviations using the **Two Standard Deviations** panel. ◊

LESSON 12

Inferences for Population Proportions

GENERAL OBJECTIVE In Lessons 8–10 we examined confidence intervals and hypothesis tests for population means. Now we will learn how to conduct those inferences for population proportions, which are themselves population means.

LESSON OUTLINE
12.1 **Confidence intervals for one population proportion**
12.2 **Hypothesis tests for one population proportion**
12.3 **Inferences for two population proportions using independent samples**

12.1 Confidence intervals for one population proportion

E-601 : I-729

Procedure 12.1 on page 728 of *IS/5e* (Procedure 11.1 on page 600 of *ES/4e*) provides a method for obtaining a large-sample confidence interval for a population proportion. Excel has all the necessary formulas for accommodating the computations involved here. We have stored these in the **1SAMP** worksheet of the Inference workbook in a panel named **One Prop**. We will use that panel to illustrate the next example.

EXAMPLE 12.1 *Illustrates the z-interval procedure in Excel*

A poll was taken of 1010 U.S. employees. The employees sampled were asked whether they "play hooky," that is, call in sick at least once a year when they simply need time to relax; 202 responded in the affirmative. Use these data to obtain a 95% confidence interval for the proportion, p, of all U.S. employees who play hooky.

SOLUTION The sample size of 1010 is certainly large enough to call for the z-interval. Typically, the data are in summary form as in this example and Excel allows for this as well as for having the original data (the yes's and no's expressed as 1's and 0's) in a column. Proceed as follows.

First, open the **DATA** worksheet of the Inference notebook and scroll to the upper left-hand corner to the column named **SD**. Then follow these steps.

Excel commands:

1 Type <u>1010</u> in input cell **B2** next to **n ==>** for the sample size n
2 Select input cell **B5**, type <u>202</u> next to **x ==>** and press [Enter⏎]

Now click on the **1SAMP** tab and locate the panel **One Prop**. We obtained the output displayed in Figure 12.1 executing the Excel commands below the figure.

FIGURE 12.1
z-Confidence interval for absentee data

z-Confidence	Interval:
95.0%	Two-sided:
Lower Limit l =	0.175
Upper Limit u =	0.225

Excel commands:

1 Type <u>SD</u> in input cell **I4** under **DATA FILE NAME:** and press [Enter⏎]
2 Select input cell **K6** and type <u>0.05</u> for α
3 Select input cell **M5** and type <u>n</u> (or anything other than y)
4 Type <u>0</u> for **Alt Code** in input cell **M6** and press [Enter⏎]

The output in the **One Prop** panel provides both a hypothesis test and a confidence interval. We will address the hypothesis test aspect in the next section of this lesson. For our present application, you will find the results recorded in the range **L7:M12**. This range is depicted in Figure 12.1. As you see, the 95% confidence interval for p runs from 0.175 to 0.225. We can be 95% confident that the proportion of all employees who play hooky is somewhere between 17.5% and 22.5%. ■

PROBLEM 12.1   A 1985 Gallup Poll estimated the support among Americans for "right to die" laws. For the survey, 1528 adults, 18 years old and older, were asked whether they were in favor of voluntary withholding of life support systems from the terminally ill. The results: 1238 said they were. Find a 99% confidence interval for the proportion of all adult Americans who are in favor of "right to die" laws.   ◊

EXACT CONFIDENCE INTERVALS

Of course, the z-interval computed this way is only approximate since the distribution of $\hat{p}$, a discrete random variable, cannot be exactly normal no matter what the value of n. For small sample sizes, it would not be advisable to use the z-procedure. For that situation, there is a more exact procedure based on the binomial distribution.

EXAMPLE 12.2   *Illustrates finding exact confidence intervals for p*

Use the data of the last example to find an exact confidence interval for the population proportion of absentees.

SOLUTION   We have included a small sample implementation in the **One Prop** panel and you may elect it by answering y in input cell **M5**. This produces more exact limits based on the binomial distribution. If you will do that, you will find the output has changed to that shown in Figure 12.2.

FIGURE 12.2
Exact confidence
interval for
absentee data

Exact-Confidence Interval:	
95.0%	Two-sided:
Lower Limit l =	0.176
Upper Limit u =	0.226

Not surprisingly, the results are quite similar to those previously obtained because the sample size is large. Nevertheless, there are differences. We recommend that you use this procedure for small sample sizes (less than about 30). ■

PROBLEM 12.2   According to genetic theory, one-fourth of the offspring of certain kinds of mice should be white, the remainder should be black or spotted. In three litters, two of these mice have produced 21 mice, 9 of which are white. Find a 95% confidence interval for the proportion of white offspring? Does it include 0.25?   ◊

12.2 Hypothesis tests for one population proportion

E-614 : I-742

As you have already observed, Excel carries out a hypothesis test as part of the **One Prop** panel. You have only to input the hypothesized proportion. To perform a large-sample hypothesis test with null hypothesis $H_0: p = p_0$, we can use the test statistic

$$z = \frac{\hat{p} - p_0}{\sqrt{p_0(1-p_0)/n}}$$

as a z-test. Let us illustrate this application using **One Prop** with Example 12.6 on page 741 in *IS/5e* (Example 11.6 on page 613 in *ES/4e*).

EXAMPLE 12.3 *Illustrates the use of Excel to test one proportion*

One of the more controversial issues in the United States is that of gun control; there are many avid proponents and opponents of banning handgun sales. In an April 1993 survey, Louis Harris of LH Research Inc. polled 1250 U.S. adults regarding their view on banning handgun sales. The results were that 650 of those sampled favored a ban. Do the data provide sufficient evidence to conclude that a majority of U.S. adults (i.e., more than 50%) favor banning handgun sales? Use the **One Prop** panel in Excel to carry out the test.

SOLUTION According to the analysis of the problem in *IS/5e* and *ES/4e*, we should perform a right tailed test of $H_0 : p = 0.5$ at the 5% level of significance. Proceed as follows.

Excel commands:

1. Open the **DATA** worksheet and type <u>1250</u> in input cell **B2** next to **n ==>** for the sample size n
2. Select input cell **B5**, type <u>650</u> next to **x ==>** and press [Enter↵]
3. Click on the **1SAMP** tab and locate the panel **One Prop**
4. Type <u>SD</u> in input cell **I4** under **DATA FILE NAME:** and press [Enter↵]
5. Select input cell **K5** and type <u>0.5</u> for p_0
6. Select input cell **K6** and type <u>0.05</u> for α
7. Select input cell **M5** and type <u>n</u> (or anything other than y)
8. Select input cell **M6** and type <u>1</u> for **Alt Code** and press [Enter↵]

Figure 12.3 on the following page displays the entire panel output. With a P-value of $0.0786 > 0.05$, there is not sufficient evidence for rejecting the null hypothesis. A lower one-sided 95% confidence bound for p is 0.497 which is further evidence for not rejecting the null hypothesis since this interval includes the value 0.5. Finally, the test statistic $z = 1.414$ does not exceed the critical value of 1.645.

FIGURE 12.3
Excel right-tailed z-test of handgun data

One Prop			
DATA FILE NAME:	n = 1250	SE CI :	0.014
SD	x = 650	phat =	0.520
	H_0: p_0 = 0.50	Exact Test ?	n
	α = 0.05	Alt Code =	1
z-Test	SE test : 0.014	z-Confidence Interval:	
Right-tailed test: z = 1.414		95.0%	One-sided:
Critical Value: 1.645		Lower Limit l =	0.497
P-value: 0.0786		p_0 =	0.50
Action: Do not Reject		Action:	Do Not Reject

Again, you may elect the more exact test for these circumstances by typing **y** in cell **M7** at Step 7 in the instructions. If you do that the output will resemble that shown in Figure 12.4.

FIGURE 12.4
Excel right-tailed Exact test of handgun data

One Prop			
DATA FILE NAME:	n = 1250		
SD	x = 650	phat =	0.520
	H_0: p_0 = 0.50	Exact Test ?	y
	α = 0.05	Alt Code =	1
Exact Test		Exact-Confidence Interval:	
Right-tailed test: x = 650.000		95.0%	One-sided:
		Lower Limit l =	0.496
P-value: 0.0829		p_0 =	0.50
Action: Do not Reject		Action:	Do Not Reject

You see that there is no change in your decision regarding the null hypothesis. ■

PROBLEM 12.3 According to the Arizona Real Estate Commission, in 1982, only 10% of the people holding real estate licenses were active in the industry. An independent agency has been asked by the Commission to determine whether this year's percentage is higher. A random sample of 150 licensed people reveals that 24 are currently active. Perform the appropriate hypothesis test at the 1% significance level. ◊

12.3 Inferences for two population proportions using independent samples
`E-625 : I-753`

Now we will examine methods for performing inferences to compare the proportions, p_1 and p_2, of two populations when we have large sample sizes for both samples and the samples are independent. Similar to the **One Prop** panel in the **1SAMP** worksheet of the **Inference** workbook, we have created a **Two Proportions** panel in the **2SAMP** worksheet in that workbook. The input is quite similar as well as the output, with both confidence intervals and tests. Data are typically found in summary form, this time two sets of data, one for each population.

The following procedures are implemented in this panel. We will combine our discussion of both hypothesis testing and confidence interval estimation.

The test of the null hypothesis

$$H_0: p_1 = p_2 \text{ (population proportions are equal)}$$

may be carried out as a z-test using the test statistic

$$z = \frac{\hat{p}_1 - \hat{p}_2}{\sqrt{\hat{p}_p(1-\hat{p}_p)}\sqrt{(1/n_1) + (1/n_2)}}$$

where

$$\hat{p}_p = \frac{x_1 + x_2}{n_1 + n_2}$$

is the pooled sample proportion, an estimate of $p = p_1 = p_2$ when H_0 is true. This is summarized as Procedure 12.3 on page 750 of *IS/5e* (Procedure 11.3 on page 622 of *ES/4e*) and is called the *Two-sample z-test*.

The endpoints of the confidence interval corresponding to the choice α are

$$(\hat{p}_1 - \hat{p}_2) \pm z_{\alpha/2} \cdot \sqrt{\hat{p}_1(1-\hat{p}_1)/n_1 + \hat{p}_2(1-\hat{p}_2)/n_2}$$

These formulas are summarized as Procedure 12.4 on page 753 of *IS/5e* (Procedure 11.4 on page 625 of *ES/4e*) and is called the *Two-sample z-interval*.

Let us implement the Excel command to solve problems involving two proportions with large sample sizes.

EXAMPLE 12.4 *Illustrates the use of the* **Two Proportions** *panel*

The U.S. National Center for Health Statistics annually conducts the National Health Interview Survey (NHIS). Information on cigarette smoking is published in *Health, United States*. The following data are based on results obtained in the NHIS.

Independent random samples of 2235 U.S. women and 2065 U.S. men were selected in order to compare the percentage of women who smoke cigarettes to the percentage of men who smoke cigarettes. Of the women sampled, 503 smoked; and of the men sampled, 572 smoked.

Do the data provide sufficient evidence to conclude that, in the United States, the percentage of women who smoke cigarettes is smaller than the percentage of men who smoke cigarettes? Test the null hypothesis at the 5% level of significance.

SOLUTION Let p_1 denote the proportion of all U.S. women who smoke cigarettes and let p_2 denote the proportion of all U.S. men who smoke cigarettes. Then the null and alternative hypotheses are

$$H_0: p_1 = p_2 \text{ (percentage of female smokers is not less)}$$
$$H_a: p_1 > p_2 \text{ (percentage of female smokers is less)}$$

Note that the hypothesis test is left-tailed since there is a less-than sign (<) in the alternative hypothesis. Proceed as follows.

Excel commands:

1. Open the **DATA** worksheet and type 2235 in input cell **B2** next to **n ==>** for the sample size n_1 and type 2065 in input cell **D2** for the sample size n_2
2. Select input cell **B5**, type 503 next to **x̄ ==>** and type 572 in input cell **D5** for x_2
3. Click on the **2SAMP** tab and locate the panel **Two Proportions**
4. Under **DATA FILE NAME:**, type SD in input cell **I4**, SD2 in input cell **I5** and press Enter
5. Select input cell **K7** and type 0.05 for α
6. Select input cell **M7**, type -1 for **Alt Code** and and press Enter
7. (Optional) If a difference other than 0 is to be tested, enter that difference in cell **M12**

Figure 12.5 displays the resulting panel output.

FIGURE 12.5 Excel output of the **Two Proportions** panel

DATA FILE NAMES:		Two Proportions		
SD	n_1 =	2235	n_2 =	2065
SD2	x_1 =	503	x_2 =	572
	$p_1$hat =	0.225	$p_2$hat =	0.277
	phat =	0.250	SE =	0.0132
	α =	0.05	Alt Code =	-1
z-Test			z-Confidence	Interval:
Left-tailed test: z =	-3.930		95%	One-sided:
Critical Value:	-1.645			
			Upper Limit u =	-0.030
P-value:	0.0000		p_1-p_2 =	0.000
Action:	Reject		Action:	Reject

A printed *P*-value of 0.0000 suggests conclusive evidence in favor of rejecting H_0. The 95% upper confidence bound on the population difference, $p_1 - p_2$, is -0.030 and the population difference 0 is NOT included in this interval. Finally, the test statistic has the value $z = -3.93$ which is less than the critical value of -1.645. Hence, the data provide sufficient evidence to conclude that, in the United States, the percentage women who smoke cigarettes is smaller than the percentage of men who smoke cigarettes. ∎

We have no exact test corresponding two the one presented in the **1SAMP** worksheet. Therefore you should be wary of applying this procedure when sample sizes are very small (say less than 20). The results, which are only approximate

12.3 Inferences for two population proportions using independent samples

because of the implementation of the normal approximation, may be misleading in such cases.

PROBLEM 12.4 The Organization for Economic Cooperation and Development, Paris, France, summarizes data on labor-force participation rates in *Labour Force Statistics*. From independent samples of 300 U.S. women and 250 Canadian women it is found that 184 of the U.S. women and 148 of the Canadian women are in their respective labor forces. At the 5% significance level, do the data suggest that there is a difference between the labor-force participation rates of U.S. and Canadian women? Find a 95% confidence interval for the difference, $p_1 - p_2$, between the labor-force participation rates of U.S. and Canadian women. ◇

LESSON 13

Chi-square Procedures

GENERAL OBJECTIVE In previous lessons, we have examined the Excel functions that are used for statistical inferences concerning population means. Now we will learn how Excel can be applied to perform *chi-square procedures*. Specifically, we will study the Excel functions required to carry out a chi-square goodness-of-fit test, and a chi-square independence test.

LESSON OUTLINE
13.1 The χ^2-distribution
13.2 **Chi-square goodness-of-fit test**
13.3 **Contingency tables; association**
13.4 **Chi-square independence test**

13.1 The χ^2-distribution

The inferential procedures discussed in this lesson rely on a class of continuous probability distributions called *chi-square distributions*. Probabilities for a random variable having a chi-square distribution are equal to areas under a curve known as a χ^2(chi-square) curve.

To perform a hypothesis test or obtain a confidence interval that is based on a chi-square distribution, we will need to be able to find the χ^2-value(s) corresponding to a specified area under a χ^2-curve. Recall that the symbol χ^2_α is used to represent the χ^2-value with area α to its right under a χ^2-curve. This quantity can be found using Table V in *IS/5e* or, alternatively, with the aid of Excel. In addition, we need to compute the area under a χ^2curve in order to compute P-values for test statistics.

FINDING THE χ^2-VALUE FOR A SPECIFIED AREA

Excel has a function called CHIINV which, with a specification of the degrees of freedom, may be used to compute the χ^2-value with a specified area to its *right*. From this, it is a simple matter to obtain the χ^2-value(s) corresponding to any specified area under an arbitrary χ^2-curve. We present the details in the example to follow.

EXAMPLE 13.1 *Illustrates how to find the χ^2-value for a specified area using Excel*

For a χ^2-curve with 20 degrees of freedom, find $\chi^2_{0.05}$; that is, find the χ^2-value with area 0.05 to its right. See Figure 13.1.

FIGURE 13.1
The χ^2-curve with 20 degrees of freedom

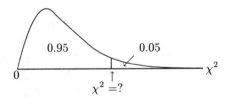

SOLUTION As we mentioned, the function CHIINV, with a specification of degrees of freedom, computes the χ^2-value with a specified area to its *right*. So, to find the χ^2value with area 0.05 to its right, simply select any empty cell in any Excel worksheet and type =CHIINV(.05,20) as a formula in that cell. You should find the value 31.4104 in the cell instantly. That is, the output shows that $\chi^2_{0.05} = 31.41042$. ∎

PROBLEM 13.1 For a χ^2-curve with 14 degrees of freedom, use Excel to find
a) the χ^2-value with area 0.10 to its right.
b) the χ^2-value with area 0.10 to it left. ◇

FINDING THE AREA FOR A GIVEN χ^2-VALUE

Using the CHIDIST function in Excel, we can find the area to the right of any specified value of χ^2. The format is CHIDIST(x,df) and the output is the area to the *right* of x under a χ^2-curve having degrees of freedom df. Hence we can find other areas as well.

EXAMPLE 13.2 *Illustrates how to find the area under a χ^2-curve for a specified value using Excel*

For a χ^2-curve with 20 degrees of freedom, find the area to the right of the value $\chi^2 = 23$. See Figure 13.2.

FIGURE 13.2
The χ^2-curve with 20 degrees of freedom

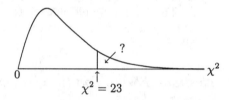

SOLUTION Select any empty cell in Excel and type =CHIDIST(23.0,20). The cell will display the value 0.2888 instantly. Hence the area to the right of $\chi^2 = 23$ for this particular χ^2-curve is approximately 0.29. ∎

PROBLEM 13.2 For a χ^2-curve with 14 degrees of freedom, use Excel to find
a) the area to the right of $\chi^2 = 13$.
b) the area between $\chi^2 = 7$ and $\chi^2 = 13$. ◇

13.2 Chi-square goodness-of-fit test E-650:I-778

The first chi-square procedure we will examine is called the *chi-square goodness-of-fit test*. Among other things, this procedure can be used to test hypotheses about the percentage distribution of a population or the probability distribution of a random variable.

13.2 Chi-square goodness-of-fit test

Excel does not have a special function to perform a chi-square goodness-of-fit test. Nonetheless, we can still use Excel to help with the computations required to carry out such a test. In the next example we will apply Procedure 13.1 on page 776 of *IS/5e* (Procedure 12.1 on page 776 of *ES/4e*) to perform a chi-square goodness-of-fit test. The calculations required will be done by Excel.

EXAMPLE 13.3 Illustrates how Excel can be used for a goodness-of-fit test

The U.S. Bureau of Investigation (FBI) compiles data on crimes and crime rates and publishes the information in *Crime in the United States*. A violent crime is classified by the FBI as murder, forcible rape, robbery, or aggravated assault. Table 13.1 provides a relative-frequency distribution for (reported) violent crimes in 1995.

TABLE 13.1 Distribution of violent crimes in the United States, 1995

Type	Relative frequency
Murder	0.012
Forcible rape	0.054
Robbery	0.323
Agg. assault	0.611
	1.000

A random sample of 500 violent-crime reports from last year yielded the frequency distribution shown in Table 13.2.

TABLE 13.2 Sample results for 500 randomly selected violent-crime reports from last year

Type	Frequency
Murder	9
Forcible rape	26
Robbery	144
Agg. assault	321
	500

At the 5% significance level, do the data provide sufficient evidence to conclude that last year's distribution of violent crimes has changed from the 1995 distribution?

SOLUTION Procedure 13.1 on page 776 of *IS/5e* (Procedure 12.1 on page 648 of *ES/4e*) has been incorporated in a panel called **GOF** in the **CHISQ** worksheet of the **Inference** workbook. Open that worksheet and locate the panel and the cell labeled **DATA**: (cell **B2**). As you see from the cells with dotted borders, there are only three data entry points. One is a column labeled **Rel Freq** for entering probabilities or percentages in decimal form. Paired with that is a column labeled **O** for entering observed frequencies. Finally, you need to enter a level of significance at which to perform the test in input cell **G4**. We have allowed for 20 entries. To implement the rest of the procedure, execute the Excel commands on the following page.

Excel commands:

1. Enter the relative frequencies (probabilities) starting in input cell **B4** and remove any data (if necessary) below your last entry
2. Enter the observed frequencies starting in input cell **C4** and remove any data (if necessary) below your last entry
3. Select cell **D4**, drag its fill handle down to the cell opposite your last entry, and remove any data (if necessary) below that cell
4. Select cell **E4**, drag its fill handle down to the cell opposite your last entry, and remove any data (if necessary) below that cell
5. Select input cell **G4** and type `0.05` for the level of significance

The necessary computed values are automatically placed in the appropriate cells (range **F5:G8**) to complete the χ^2 test as displayed in Figure 13.3.

FIGURE 13.3 Excel goodness-of-fit test of the crime data

DATA:				GOF	
Rel Freq	O	E	(O-E)²/E	df=	3
0.012	9	6	1.5	α=	0.05
0.054	26	27	0.037037	χ^2 =	4.220
0.323	144	161.5	1.896285	χ^2_α =	7.815
0.611	321	305.5	0.786416	P-value :	0.239
				Action :	Do not Reject

Before you examine the results, you need to check a couple of conditions, namely, the *ASSUMPTIONS* that need to be satisfied. They are found in Procedure 13.1 on page 776 of *IS/5e* (Procedure 12.1 on page 648 of *ES/4e*).

1. By examining the expected frequency column (labeled E) you see that all expected frequencies are at least 1.

2. At most 20% of the expected frequencies are less than 5 (since there are no expected frequencies less than 5).

Thus, the assumptions are satisfied.

In cell **G5** you see that the value of the test statistic is $\chi^2 = 4.220$. Comparing this with the critical value of $\chi^2_{0.05} = 7.815$ located in cell **G6** and recalling that the test is right-tailed, we see that the value of the test statistic does *not* fall in the rejection region. The *P*-value appears in cell **G7** and has the value 0.239, considerably greater than the level of significance, 0.05.

Either way, we do not reject the null hypothesis and are reminded of this action in cell **G8**. In other words, the sample data do not provide sufficient evidence to conclude that last year's violent-crime distribution has changed from the 1995 distribution. ■

PROBLEM 13.3 According to the document *Current Housing Reports,* published by the U.S. Bureau of the Census, the primary-heating-fuel distribution for occupied housing units is as follows:

Primary heating fuel	Percent
Natural gas	56.7
Fuel oil, kerosene	14.3
Electricity	16.0
LPG	4.5
Wood	6.7
Other	1.8

A random sample of 250 occupied housing units built after 1974 yields the frequency distribution below.

Primary heating fuel	Frequency
Natural gas	91
Fuel oil, kerosene	16
Electricity	110
LPG	14
Wood	17
Other	2

Do the data provide sufficient evidence to conclude that the primary-heating-fuel distribution for occupied housing units built after 1974 differs from that of all occupied housing units? Use $\alpha = 0.05$. Use Excel to test the appropriate hypothesis, as in Example 13.3. ◇

13.3 Contingency tables; association

Sometimes surveys are initiated in order to determine whether or not there is a relationship between two variables that are classified in two ways. Indeed, many times the observed values of these variables are not even numerical. Yet we can determine the frequency of occurrence in each classification and produce a two-way table called a *contingency table*. Such tables now have numerical entries and can then be analyzed by statistical methods.

CONTINGENCY TABLES E-658 : I-786

The raw data for a contingency analysis may be in the form of simply a tabulation of responses to a survey or a classification into two or more nonnumeric categories. You may find this type of data in many sources including the Internet. The data can be input directly into a Excel worksheet to produce a contingency table and proceed to the analysis. Of course such data usually contain hundreds if not thousands of observations and hence we look to the computer for converting the raw data into a contingency table. In the case of very large sets of data that are available in a format for direct input to the computer, consult your reference on the various ways of importing data into the Excel worksheet. Here is an example.

EXAMPLE 13.4 *Responses by sex to a questionnaire*

The data to follow are an abbreviated (for the sake of convenience) table of responses to a survey asking respondents whether or not they approve of the way Congress currently carries out its business. Respondents were only allowed to answer with Yes, No, or Und (for undecided). The results appear in the Table 13.3. They are arranged in 40 pairs as they were collected, each pair corresponding to the sex of the respondent followed by that respondent's reply to the question.

TABLE 13.3
Survey on Congressional approval

Sex	Reply	Sex	Reply
Male	Yes	Female	No
Male	No	Male	No
Female	Yes	Male	No
Male	Und	Female	Yes
Male	No	Male	Und
Female	Yes	Female	No
Female	Und	Female	Yes
Male	Yes	Male	No
Male	No	Male	Und
Male	Yes	Male	Und
Female	No	Male	No
Male	Yes	Male	No
Male	No	Male	No
Female	No	Male	Und
Male	No	Male	Und
Male	Yes	Female	No
Female	No	Female	Und
Female	Yes	Female	Yes
Female	Yes	Female	No
Female	Und	Female	Und

Convert these data into a two-way contingency table using Excel.

SOLUTION To produce the contingency table, we entered the data from Table 13.3 into the **CHISQ** worksheet in three neighboring columns named **SEX**, **REPLY** and **FREQ**. The frequency column has an entry of 1 in each cell since each separate occurrence is given so. This means that we can use Excel's capability of entering patterned data to advantage. Sometimes you will encounter a contingency table where the frequency of occurrence of each category is already summarized (see the next section). In that case, you need to enter the data as triplets with the **FREQ** entry equal to the number of times each combination occurred. Proceed as follows.

13.3 Contingency tables; association

Excel commands:

1. Starting in cell **J2**, enter the pair (Male, Yes) in the pair of cells (**J2, K2**) for row 2 of the table
2. Enter the pair (Male, No) in the pair of cells (**J3, K3**) for row 3 of the table
3. Continue entering pairs this way until you enter (Female, Und) for the last row (row 41) of the table
4. Select cell **L1**, type 1, and then drag the cell fill handle down to cell **L41**

With the data now located in those cells, proceed as follows to construct and display the contingency table.

Excel commands:

1. Select **D**ata ▸ **P**ivot Table Report...
2. Select the **M**icrosoft **E**xcel list or data base option and click [Next >]
3. Type J1:L41 in the **R**ange: text box (or use the [▦] button to select the range) and click [Next >]
4. Drag the [SEX] button into the **R**OW box, drag the [REPLY] button into the **C**OLUMN box, drag the [FREQ] button into the **D**ATA box, and press [Enter ↵]
5. Select the **E**xisting worksheet option, and enter any cell as the corner of a block of empty cells (we chose **N3**)
6. Select [Finish]

You should see a table resembling Figure 13.4. But the order of columns and rows is different since Excel alphabetizes the names. We changed our table to have Yes replies first by selecting cell **P4** and typing Yes.

FIGURE 13.4
Excel contingency table

Sum of FREQ	REPLY			
SEX	Yes	No	Und	Grand Total
Female	7	7	4	18
Male	5	11	6	22
Grand Total	12	18	10	40

Totals are labeled **Grand Total** by Excel and you see the resemblance of this table to the Table 13.9 on page 785 of *IS/5e* (Table 12.9 on page 257 of *ES/4e*). Each cell contains the frequency of respondents in the joint classification of the row and column marking that cell. Row totals give summaries of the number of respondents in each Sex category, while column totals summarize by the Reply category. ■

PROBLEM 13.4 Repeat the construction of a contingency table for Example 13.4 using the data of Figure 13.4 as 6 triplets instead of 40 as in the example. ◊

ASSOCIATION

E-660 : I-788

To examine the association between Sex and Reply in the population, we need to construct and examine a conditional probability distribution as defined and discussed on pages 786, 787 in *IS/5e* (pages 658, 659 in *ES/4e*).

EXAMPLE 13.5 *Association between Sex and Reply*

Refer to Example 13.4. Use Excel to examine the relationship between the two variables Sex and Reply.

SOLUTION We used the following Excel commands.

Excel commands:

1. Select **D**ata ➤ **P**ivot Table Report...
2. Select the **M**icrosoft Excel list or data base option and click [Next>]
3. Type **J1:L41** in the **R**ange: text box (or use the button to select the range) and click [Next>]
4. Drag the [SEX] button into the **R**OW box, drag the [REPLY] button into the **C**OLUMN box, drag the [FREQ] button into the **D**ATA box, and click [Next>]
5. Double click on **Sum of FREQ** field and type `Conditioned on REPLY` in the **Na**me: text box
6. Select [Options>>], click the button in the Show d**a**ta as: text box and select **% of column**
7. Select [Number...], select **Number** from the drop down menu and select **3** for the number of decimal places
8. Click [OK] twice, select an output cell (we chose **N9**) and then select [**F**inish]

The output (range **N9:R13**) is shown in Figure 13.5.

FIGURE 13.5
Excel conditional (on REPLY) probability table

Conditional Distribution	REPLY			
SEX	Yes	No	Und	Grand Total
Female	0.583	0.389	0.400	0.450
Male	0.417	0.611	0.600	0.550
Grand Total	1.000	1.000	1.000	1.000

For example, there is a 45% chance that a respondent is female but if we knew the response to the questionnaire was Und, there is only a 40% chance that it came from a female. There appears to be an association between these two variables. ∎

PROBLEM 13.5 Using the outputs of this section, practice drawing your own conclusions for various choices of variables. *(Note:* You may wish to use a separate worksheet for your own practice.) ◇

13.4 Chi-square independence test

[E-673 : I-801]

In the last section, we restricted our conclusions to the hypothetical condition that the data represented a population. However, in almost all cases, these surveys represent a sample of the entire population, since data for the entire population are usually not available. Thus, in most cases, we must employ inferential statistics to decide whether two characteristics of a population are statistically dependent.

One of the most commonly used tests for statistical dependence is the *chi-square independence test*. The steps for carrying out this test are given in Procedure 13.2 on page 800 of *IS/5e* (Procedure 12.2 on page 672 of *ES/4e*). The computations required are generally tedious and time-consuming. Fortunately, the Excel functions will allow us to implement the steps of the procedure when data are available in a contingency table. While the format is a bit cumbersome, we will carry out the following example as a general model for you to use in a new situation. You will have to adjust ranges accordingly, but we will indicate where this must be done.

EXAMPLE 13.6 Illustrates the use of Excel to test independence

A national survey was conducted to obtain information on the alcohol consumption patterns of American adults by marital status. A random sample of 1772 residents, 18 years old and over, yielded the data shown in Table 13.4.[†]

TABLE 13.4
Contingency table of marital status versus alcohol consumption for 1772 randomly selected adults

	DRINKS PER MONTH			
	Abstain	1–60	Over 60	Total
Single	67	213	74	354
Married	411	633	129	1173
Widowed	85	51	7	143
Divorced	27	60	15	102
Total	590	957	225	1772

STATUS

Do the data suggest, at the 5% significance level, that marital status and alcohol consumption patterns are statistically dependent? As pointed out in *IS/5e* and *ES/4e* the null hypothesis is that marital status and alcohol consumption are statistically *independent*. Rejection of that hypothesis leads to the conclusion that there is a relationship between marital status and alcohol consumption.

[†] Adapted from: Clark, W. B. and Midanik, L. "Alcohol Use and Alcohol Problems among U.S. Adults: Results of the 1979 National Survey." In: National Institute on Alcohol Abuse and Alcoholism. Alcohol and Health Monograph No 1, Alcohol Consumption and Related Problems. DHHS Pub. No (ADM) 82–1190, 1982.

SOLUTION We want to perform the hypothesis test

H_0: Marital status and alcohol consumption are statistically independent.

H_a: Marital status and alcohol consumption are statistically dependent.

at the 5% significance level.

To employ Excel, we first entered the data into the computer in three columns named Marital, Drinks/mo, and Freq in the **CHISQ** worksheet of the **Inference** notebook.

Next, we created the contingency table with the **Pivot Table Report...** option of the **Data** menu in Excel just as in the last section. You will find that output in the range **X3:AB9**. See Figure 13.6 for the output and compare the results with Table 13.13 on page 796 of *IS/5e* (Table 12.13 page 668 of *ES/4e*).

FIGURE 13.6 Excel contingency table for marital status vs. drinks per month

Sum of Freq	Drinks/mo			
Marital	Abstain	1 to 60	Over 60	Grand Total
Single	67	213	74	354
Married	411	633	129	1173
Widowed	85	51	7	143
Divorced	27	60	15	102
Grand Total	590	957	225	1772

Then we created the Expected Frequency table in the range **AC5:AE8** by implementing the formula $E = (R \cdot C)/n$, located on page 799 of *IS/5e* (page 672 of *ES/4e*), for each cell using the following commands.

Excel commands:

1 Select cell **AC5** and type `=Y9*$AB5/$AB$9`
2 Drag the fill handle of cell **AC5** to the right over the range **AD5:AE5**
3 Repeat the last two instructions for the rows **AC6:AC8** incrementing only the **$AB**x location from $x = 6$ through $x = 8$

This table is reproduced for you in Figure 13.7.

FIGURE 13.7 Excel expected frequency table for marital status vs. drinks per month

Expected Frequencies		
117.87	191.18	44.95
390.56	633.50	148.94
47.61	77.23	18.16
33.96	55.09	12.95

You will find that the entries in this table match the expected frequencies entered in Table 13.14 of *IS/5e* on page 798 (Table 12.14 on page 670 of *ES/4e*).

Following this we created a table to account for each cell contribution to the value of the χ^2 test statistic in the range **AC10:AE13**. This was accomplished by first typing the formula =(Y5-AC5)^2/AC5 in cell **AC10**. Then drag the fill handle of that cell over the rest of the range. The result is displayed in Figure 13.8. Notice that these entries match the individual values for Chi-Sq listed after Table 13.7 on page 785 of *IS/5e* (Table 12.14 on page 670 of *ES/4e*).

FIGURE 13.8
Individual cell contributions

Individual	Cell Contribution to ==>	
21.95	2.49	18.78
1.07	0.00	2.67
29.36	8.91	6.86
1.43	0.44	0.32

Finally, the implementation of the Chi-square test for independence is carried out in the range **AF4:AG10**, producing the result shown in Figure 13.9.

FIGURE 13.9
The chi-square test for independence

χ^2 Test for Independence	
α =	0.05
df =	6
P-value =	0.000
χ^2_α =	12.592
χ^2 =	94.269
Action	Reject

For the record, here is what we entered in the various cells.

Excel commands:

1 Select input cell **AG5** and type .05
2 Select cell **AG6** and type the formula =(COUNT(Y5:Y8)-1)*(COUNT(Y5:AA5)-1) for degrees of freedom (This range must be adjusted for a different table)
3 Select cell **AG7** and type the formula =CHIDIST(AG9,AG6) (These cells must be adjusted for a different table)
4 Select cell **AG8** and type the formula =CHIINV(AG5,AG6) (These cells must be adjusted for a different table)
5 Select cell **AG9** and type the formula =SUM(AC10:AC13,AD10:AD13,AE10:AE13) (These ranges must be adjusted for a different table)
6 Select cell **AG10** and type the formula =IF(AG7<AG5,Reject,Do not reject) (These cells must be adjusted for a different table)

148 Lesson 13 Chi-square Procedures

The results confirm the analysis given in *IS/5e* and *ES/4e*. You are reminded to reject the null hypothesis with strong evidence against it in the form of both a P-value of practically 0 and such a large χ^2 value compared to the critical value for a right-tailed test. We conclude that there is an association between marital status and alcohol consumption. ∎

Not a particularly elegant solution with a lot of precise details to implement. Certainly you should practice your own solutions on a different worksheet using this solution as a model and adjust cell names accordingly.

PROBLEM 13.6 Use the data and contingency table results for Example 13.4 on page 142 to conduct a Chi-square test for independence of responses and sex using Excel. Draw the appropriate conclusion at the 5% level of significance. ◊

LESSON 14

Descriptive Methods in Regression and Correlation

GENERAL OBJECTIVE It is frequently of interest to know whether two or more variables are related and, if so, how they are related. For instance, is there a relationship between college GPA and SAT scores? If these variables are related, how are they related? The president of a large corporation knows there is a tendency for sales to increase as advertising expenditures increase. But how strong is that tendency and how can she predict the approximate sales that will result from various advertising expenditures?

Some commonly used methods for examining the relationship between two or more variables and for making predictions are provided by *linear regression and correlation*. Descriptive methods in linear regression and correlation will be addressed in this lesson. Inferential methods in linear regression and correlation will be addressed in the next lesson.

LESSON OUTLINE
14.1 The regression equation
14.2 The coefficient of determination
14.3 Linear correlation

14.1 The regression equation

In this section we will learn how Excel can be applied to determine the regression line for a set of data points. Recall that the *regression line* is the best-fitting line to a set of data points, best-fitting according to the *least-squares criterion*.

SCATTER DIAGRAMS

E-204 : I-828

The idea behind finding a regression line is based on the assumption that the data points are actually scattered about a straight line. Consequently, before determining the regression line, we need to look at a *scatter diagram* of the data. We should then proceed with the regression only if the data points appear to be scattered about a straight line.

Excel has the capability of producing scatter diagrams once the data are appropriately stored in the workspace. We illustrate in the following example.

EXAMPLE 14.1 *Illustrates creating a scatter diagram*

Table 14.1 displays data on age and price for a sample of cars of a particular make and model. We will refer to the car as the Orion automobile, even though the data were gathered for a real car. The data were obtained from the *Asian Import* edition of the *Auto Trader* magazine. Ages are in years; prices are in hundreds of dollars, rounded to the nearest hundred dollars. The goal is to predict price from age.

TABLE 14.1
Age - price data for Orions

Car	Age (yrs) x	Price ($100s) y
1	5	85
2	4	103
3	6	70
4	5	82
5	5	89
6	5	98
7	6	66
8	6	95
9	2	169
10	7	70
11	7	48

Our first task is to plot the data, price versus age, in order to detect any possible relationship between age and price, keeping in mind that the data only represent a sample of all Orions. Use Excel to obtain a scatter diagram of these data.

SOLUTION We have entered the data in the **DATA** worksheet of the Regression workbook under the names AGE and PRICE. To obtain a scatter diagram of the age-price data, open up the **REGRESS** worksheet in the Regression workbook and proceed as follows.

Excel commands:

1. Click the chart wizard button [⊞] in the Excel toolbar
2. Choose **XY (Scatter)** as th **Chart type:** and click [Next >]
3. Select the **Data range:** text box and type `PRICE:AGE`
4. Select the **Columns** radio button and click [Next >]
5. In the **Chart Options** dialogue box, type `Scatter Diagram` in the **Chart title:** text box
6. Select the **Value (X) axis:** text box and type `AGE`
7. Select the **Value (Y) axis:** text box and type `PRICE`
8. Click [Next >] and deselect the **Show Legend** check box in the **Legend** folder
9. Click [Next >] and select the **As object in:** radio button, type `REGRESS` in the text box and select [Finish]

Figure 14.1 shows the resulting output relocated to the range **Q23:S36** in the **REGRESS** worksheet. (We deleted the legend since we already have the information needed.)

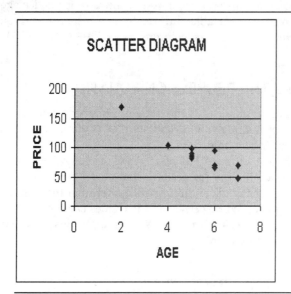

FIGURE 14.1
Excel scatter diagram for the Orion data

Note that the variable corresponding to the first storage location, in this case PRICE, is plotted on the vertical axis and that the variable corresponding to the second storage location, in this case AGE, is plotted on the horizontal axis. Since AGE in this case is the independent variable, we call it the *predictor variable* in regression analysis. Correspondingly, PRICE, the dependent variable, is referred to as the *response variable*. Again, the goal is to predict price from age.

It does appear that the data points are scattered about a straight line. Therefore, it is reasonable to find a regression line for that data. ∎

PROBLEM 14.1 The U.S. National Center for Health Statistics collects and publishes data on heights and weights by age and sex. A random sample of 11 males, aged 18–24 years, gave the data in the following table.

Height (in) x	Weight (lb) y
65	175
67	133
71	185
71	163
66	126
75	198
67	153
70	163
71	159
69	151
69	155

Use Excel to obtain a scatter diagram of the data for males aged 18–24 years using height as a predictor variable and weight as a response variable.

We have stored the data in columns named HEIGHT and WEIGHT in the **DATA** worksheet. ◊

OBTAINING THE REGRESSION EQUATION   E-209 : I-833

As we found in Example 14.1, the data on price versus age for Orions appears to be scattered about a straight line. Therefore, it is reasonable to proceed to determine the regression line for that data.

Formula 14.1 on page 831 of *IS/5e* (Formula 4.1 on page 207 of *ES/4e*) allows us to determine the regression equation for a set of data points. Excel has a program that, among other things, will determine the regression equation for us. That program is called **Regression** and it resides as an option in Analysis Tools under **Data Analysis...** in the **Tools** menu. We will apply this tool to obtain the regression equation for the Orion data.

EXAMPLE 14.2 *Illustrates finding a regression equation*

The data on age and price for a sample of Orions are given in Table 14.1 on page 150. Use Excel to obtain the regression equation for that data.

SOLUTION Open up the **REGRESS** worksheet and proceed as follows.

Excel commands:

1 Choose **Tools ➤ Data Analysis...** and select **Regression** from among the tools options

2 Select the **Input Y Range:** text box and type PRICE (or enter the range **DATA!B1:B12**)

3. Select the **Input X Range:** text box and type AGE (or enter the range **DATA!A1:A12**)
4. Select the **Labels** check box
5. Select the **Output range:** button and enter **A2** in its text box
6. click [OK]

The Regression function tells Excel to perform a regression analysis on the age and price data with AGE as the single independent (predictor) variable and PRICE the dependent (response) variable. You will find the output displayed in the range **G3:O20** on the worksheet (with some additional formatting for number of decimal places). Figure 14.2 displays the results we are currently interested in, found in the range **A17:B19**.

FIGURE 14.2
Partial Excel regression output for the Orion data

	Coefficients
Intercept	195.468
AGE	-20.261

These two constants (*Coefficients*) allow us to write the required regression equation:

PRICE = 195 − 20.3 AGE

In other words, the regression equation is $\hat{y} = 195 - 20.3x$.

We will discuss several of the other aspects of the regression output in future sections. ∎

PROBLEM 14.2 Refer to Problem 14.1 on page 152. Use Excel to obtain the regression equation of the weight-versus-height data, using WEIGHT as the response variable and HEIGHT as the predictor. ◇

14.2 The coefficient of determination [E-224 : I-848]

Section 14.3 of *IS/5e* (Section 4.3 of *ES/4e*) discusses the *coefficient of determination*, r^2. The coefficient of determination is a descriptive measure of the utility of the regression line for making predictions. Specifically, r^2 gives the percentage reduction in the total squared error obtained by using the regression equation to predict the observed y-values, instead of just predicting the mean $\overline{y}$.

The coefficient of determination, r^2, is defined in terms of the *total sum of squares*, $SST = \Sigma(y - \overline{y})^2$, and the *error sum of squares*, $SSE = \Sigma(y - \hat{y})^2$. We have

$$r^2 = 1 - \frac{SSE}{SST}$$

Example 14.7 of *IS/5e* beginning on page 845 of *IS/5e* (Example 4.7 beginning on page 221 of *ES/4e*) traces the calculations required to find the coefficient of determination of the age-versus-price data for Orions. As we saw, those calculations are extensive. Excel will do all of that work for us. In fact, the output of the regression function, which we used to obtain the regression equation, also gives the value of the coefficient of determination.

EXAMPLE 14.3 *Illustrates finding the coefficient of determination*

Use Excel to find the coefficient of determination for the circumstances of Example 14.2.

SOLUTION The regression output in the **Regression worksheet** currently of interest is the range **A4:B9** depicted in Figure 14.3.

FIGURE 14.3
Partial Excel regression output for the Orion data

Regression Statistics	
Multiple R	0.924
R Square	0.853
Adjusted R Square	0.837
Standard Error	12.577
Observations	11

The second item listed below the title *Regression Statistics* as **R Square** shows the value as 0.853. That is, $r^2 = 85.3\%$. ■

PROBLEM 14.3 Refer to Problem 14.2 on page 153. Use the Excel output you obtained in that problem to determine the coefficient of determination, r^2. ◇

EXPLAINED VARIATION

E-227 : I-851

There is another way to interpret the coefficient of determination, r^2. Namely, it can be interpreted as the percentage of variation in the observed y-values that is explained by the regression. The justification for this interpretation is given on pages 848, 849 of *IS/5e* (pages 225, 226 in *ES/4e*). We briefly discuss it here.

The deviation, $y - \overline{y}$, of an observed y-value from the mean can be decomposed into two parts—the deviation that is explained by the regression line, $\hat{y} - \overline{y}$, and the remaining unexplained deviation (error), $y - \hat{y}$. Therefore, the total amount of variation (squared deviation) explained by the regression line is $\Sigma(\hat{y} - \overline{y})^2$. This is called the *regression sum of squares*, and is denoted **SSR**. Hence,

$$SSR = \Sigma(\hat{y} - \overline{y})^2$$

The total variation in the observed y-values is given by the total sum of squares, $SST = \Sigma(y - \overline{y})^2$. Consequently, the percentage of the total variation in the observed

14.2 The coefficient of determination

y-values that is explained by the regression is SSR/SST. It can be shown that this last quantity is equal to the coefficient of determination, r^2 and it is always true that $SSR + SSE = SST$ (the regression identity).

Excel can be quite helpful in determining these various sums of squares. To obtain the three sums of squares, we use the table in the output of the Regression option, entitled ANOVA.

EXAMPLE 14.4 *Illustrates computing the coefficient of determination from sum of squares*

Use the ANOVA output in Excel to identify the various sums of squares for the price-versus-age data of the Orions. Use these results to interpret the coefficient of determination.

SOLUTION The items of interest for this example may be found in the range **A11:F15** as displayed in Figure 14.4.

FIGURE 14.4 Partial Excel regression output for the Orion data

ANOVA	df	SS	MS	F	Significance F
Regression	1	8285.014	8285.014	52.380	0.000
Residual	9	1423.532	158.170		
Total	10	9708.545			

From Figure 14.4, the first entry in the SS column is the regression sum of squares, SSR, the second is the error sum of squares, SSE, (called Residual by Excel), and the third is the total sum of squares, SST. So, we see that $SSR = 8285.0$, $SSE = 1423.5$, and $SST = 9708.5$. As expected, $SST = SSR + SSE$.

The percentage of the total variation in the observed y-values that is explained by the regression is

$$\frac{SSR}{SST} = \frac{8285.0}{9708.5} = 0.853$$

or 85.3%. In other words, about 85% of SST is equal to the regression sum of squares SSR. This is the sense in which we interpret the phrase "explained by regression." But, as noted earlier, this quantity is the coefficient of determination, r^2 printed in the Regression Statistics portion of the output. ∎

PROBLEM 14.4 Refer to Problem 14.2 (page 153) and Problem 14.3 (page 154). Use the Excel output you obtained in Problem 14.2 to find the regression sum of squares SSR, the error sum of squares SSE, and the total sum of squares, SST. Determine the coefficient of determination from these sums of squares. ◇

14.3 Linear correlation

E-234:I-858

It is often of interest to know whether there is an association or *correlation* between two variables. For example, is there a positive correlation between advertising expenditures and sales? Or, are IQ and alcohol consumption uncorrelated?

One of the most common measures of the correlation between two variables is the *linear correlation coefficient*, r. The linear correlation coefficient is a single number that can be used to describe the strength of the *linear* (straight-line) relationship between two variables.

The linear correlation coefficient, r, is always between -1 and 1. Positive values of r suggest that the variables x and y are *positively linearly correlated*, meaning that y tends to increase linearly as x increases, with the tendency being greater the closer that r is to 1. Negative values of r suggest that the variables are *negatively linearly correlated*, meaning that y tends to decrease linearly as x increases, with the tendency being greater the closer that r is to -1. Values of r close to 0 suggest that the variables are *linearly uncorrelated*, meaning that there is neither a tendency for y to increase as x increases nor a tendency for y to decrease as x increases—in a *linear* fashion.

The linear correlation coefficient, r, for n data points can be computed using the formula

$$r = \frac{n(\Sigma xy) - (\Sigma x)(\Sigma y)}{\sqrt{n(\Sigma x^2) - (\Sigma x)^2}\sqrt{n(\Sigma y^2) - (\Sigma y)^2}}$$

Excel has a routine that will compute r for us. The next example illustrates this.

EXAMPLE 14.5 *Illustrates correlation*

The data on age and price for a sample of 11 Orions are displayed in the second and third columns of Table 14.1 on page 150. Use Excel to determine the linear correlation coefficient of the data.

SOLUTION Recall that we have previously stored the age and price data in columns named AGE and PRICE. To find the linear correlation between the two variables, find any empty cell in Excel and type =CORREL(AGE,PRICE). That cell will instantly display the value -0.92378 depending on the number format of the cell.

Evidently, there is a strong negative linear correlation between age and price of Orions. This means that as the age of a Orion increases, the price tends to decrease in a linear fashion. Incidentally, it is no coincidence that the coefficient of determination is denoted r^2, for the square of the correlation coefficient is indeed the coefficient of determination. ∎

PROBLEM 14.5 Use the weight-versus-height data for males 18–24 years old in Problem 14.1, page 152, to find the linear correlation coefficient of HEIGHT and WEIGHT. ◇

LESSON 15

Inferential Methods in Regression and Correlation

GENERAL OBJECTIVE We often need to know whether a relationship exists between two variables, and if so, what that relationship is. For example, is there a relationship between SAT score and college grade-point average? If these variables are related, how are they related? Linear regression analysis and correlation analysis provide some of the most widely used procedures for examining the relationship between two variables. In this lesson we will learn how Excel can be applied to perform linear regression and correlation analyses.

LESSON OUTLINE
15.1 The regression model:analysis of residuals
15.2 Inferences for the slope of the population regression line
15.3 Estimation and prediction
15.4 Inferences in correlation
15.5 Testing for normality

15.1 The regression model: analysis of residuals

Up to now we have only examined *descriptive* methods in linear regression and correlation. In order to make *inferences* in linear regression and correlation, it is necessary for the variables under consideration to satisfy certain conditions. We list them below.

Assumptions for regression inferences

1. *Population regression line:* There are constants β_0 and β_1 such that for each value x of the predictor variable X, the conditional mean of the response variable Y is $\beta_0 + \beta_1 x$. We refer to the straight line $y = \beta_0 + \beta_1 x$ as the *population regression line* and to its equation as the *population regression equation*.
2. *Equal standard deviations:* The conditional standard deviations of the response variable Y are the same for all values of the predictor variable X. We denote this common standard deviation by σ.
3. *Normality:* For each value x of the predictor variable X, the conditional distribution of the response variable Y is a normal distribution.
4. *Independent observations:* The observations of the response variable are independent of one another.

In other words, Assumptions (1)–(3) require that there exist constants β_0, β_1, and σ so that for each value x of the predictor variable X, the conditional distribution of the response variable, Y, is a normal distribution having mean $\beta_0 + \beta_1 x$ and standard deviation σ. These assumptions are often referred to as the *regression model*.

THE STANDARD ERROR OF THE ESTIMATE   E-727 : I-877

Let us suppose that we are considering two random variables, X and Y, satisfying the regression model. The parameters β_0, β_1, and σ, are almost always unknown and, hence, must be estimated from the sample data. The statistic used to estimate σ is called the *standard error of the estimate* or the *residual standard deviation* and is defined as follows:

$$s_e = \sqrt{\frac{SSE}{n-2}}$$

where, as we know, $SSE = \Sigma(y - \hat{y})^2$.

In Section 14.2, we learned that Excel can be used to obtain the (sample) regression equation for a set of data points. The output resulting from the application of the **Regression** tool displays the value of the standard error of the estimate, s_e, (labeled **Standard Error** on the output) as well as the regression equation and several other items of interest. There are also many other options available in the regression tool that we did not take advantage of when we used it in the last chapter. Thus, we will produce the output again (overwriting the previous one) and selecting other choices available. We illustrate this in the next example.

15.1 The regression model: analysis of residuals

EXAMPLE 15.1 *Illustrates finding the standard error of the estimate*

Recall the Excel regression output applied to the data on age and price for a sample of 11 Orions located in the **REGRESS** worksheet of the Regression workbook. The summary statistics in the range **A2:B9** are duplicated in Figure 15.1.

FIGURE 15.1
Partial Excel regression output for the Orion data

SUMMARY OUTPUT

Regression Statistics	
Multiple R	0.924
R Square	0.853
Adjusted R Square	0.837
Standard Error	12.577
Observations	11

Use the output in Figure 15.1 to determine the standard error of the estimate for the sample of 11 Orions.

SOLUTION The standard error of the estimate, s_e, is the entry in the fourth line of the output labeled **Standard Error**, namely, 12.577. Thus, for the Orion data, we may state that $s_e = 12.577$. ∎

PROBLEM 15.1 Refer to the Excel output you obtained in Problem 14.2 on page 153.
a) What is s_e for the height and weight data?
b) Presuming that the random variables height, X, and weight, Y, for males 18–24 years old satisfy the assumptions for regression inferences, what is your estimate of the common population standard deviation, σ, of weights for males 18–24 years old of any given height? ◇

ANALYSIS OF RESIDUALS   [E-731 : I-881]

We can also use the regression routine to obtain residuals and then plot the results, plotting the values of the residuals obtained by considering the predicted price against the values of **Age**. We can then examine the residual plot for lack of pattern. To the extent that plot has little if any discernible pattern, we would feel comfortable about having satisfied some of the regression assumptions. We should also examine the normal probability plot of the residuals for linearity as further confirmation of the regression assumptions. We illustrate this in the next example.

EXAMPLE 15.2 *Illustrates residuals and residual plots*

Using the data of the preceding example, obtain the residuals, residual plot, and a normal probability plot of the residuals for the age and price data of the Orions.

160 Lesson 15 Inferential Methods in Regression and Correlation

SOLUTION With the data still stored in columns named PRICE and AGE, open up the REGRESS worksheet and proceed as follows.

Excel commands:

1 Choose Tools ➤ Data Analysis... and select Regression from among the tools options
2 Select the Input Y Range: text box and type PRICE
3 Select the Input X Range: text box and type AGE
4 Select the Labels check box
5 Click the Output range: button and enter A2 in its text box
6 Select the Residuals and Residual Plots check boxes
7 Click [OK]

Next, we carried out some formatting as follows.

Excel commands:

1 Locate the AGE Residual Plot, select the plot and move it to cover the range H22:L35
2 Double click anywhere in the Value (Y) axis, format the axis for number with 1 decimal place, click the Scale folder tab and type -15 in the Value (X) axis Crosses at: text box
3 Choose View ➤ Toolbars and select the Drawing check box
4 Click on the Line button in the Drawing toolbar, point to 0.0 on the Value (Y) axis, drag a straight line across the output box and right click the box at one end of the line
5 Select Format AutoShape... and among the Dashed: options available, select Round Dot (the second choice)
6 click [OK]

The results are displayed in Figure 15.2.

FIGURE 15.2
Excel residual plot

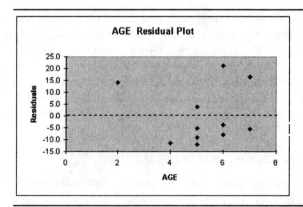

15.1 The regression model: analysis of residuals

The residuals fall roughly in a horizontal band centered about 0 with no clear pattern on the **Age** axis so we are satisfied with the regression assumptions. The value at 2 is a little bothersome but not enough to reject the validity of performing linear regression.

We can also obtain a normal probability plot for the residuals by first converting them to Nscores and then plotting those Nscores against the residuals themselves. Although there is a probability plot option in the regression tool, we will use the methods of Lesson 6 for the sake of consistency. Proceed as follows.

Excel commands:

1. In the RESIDUAL OUTPUT range (**A23:C36**) in the **REGRESS** worksheet, type `RESI` in cell **D25** and type `NSCORE` in cell **E25**
2. Copy the range **C26:C36** to the range **D26:D36** using the copy and paste buttons in the toolbar
3. Select any filled cell in column **D** and click the sort button in the toolbar
4. Select cell **E26** and type `=NORMSINV((A26-3/8)/($A$36+1/4))`
5. Drag the fill handle in **E26** through cell **E36**
6. Select the range **D26:E36**
7. Click the chart wizard button in the toolbar and select **XY (Scatter)** as the chart type
8. Click `Next >` twice
9. In the Titles folder type `Probability Plot` in the **C**hart title: text box
10. Select the **V**alue(X) Axis text box and type `Residual`
11. Select the **V**alue(Y) Axis text box and type `NSCORE`
12. Select `Finish`

We produced the chart shown in Figure 15.3 in the **Regression** worksheet but only after executing the Excel formatting commands that occur after the figure on the following page.

FIGURE 15.3
Excel normal probability plot for residuals

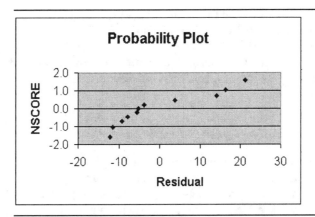

Excel commands:

1. Locate the Probability Plot, select the plot and move it to cover the range **H38:L51**
2. Select the caption [♦ *Series 1*] and click the [Del] key
3. Double click anywhere in the **Value (Y) axis**, format the axis for number with 1 decimal place, click the Scale folder tab and type **-2** in the **Value (X) axis Crosses at:** text box
4. Double click anywhere in the **Value (X) axis**, format the axis for number with 0 decimal places
5. Click the Scale folder tab and type **-20** in the **Value (Y) axis Crosses at:** text box

The normal probability plot in Figure 15.3 is roughly linear, again taking into account the small sample size. At least the departure from linearity is not so extreme that we would consider our assumption of normality violated. ∎

PROBLEM 15.2 Refer to Problem 14.1 on page 152. Using Excel, obtain the residuals, residual plot, and a normal probability plot of the residuals for the regression of weight on height.

15.2 Inferences for the slope of the population regression line

In this section, we will learn how Excel can be used to perform the inferential procedures discussed in Section 15.2 of *IS/5e* (Section 14.2 of *ES/4e*). That is, we will examine inferences concerning β_1—the slope of the population regression line.

HYPOTHESIS TESTS FOR THE SLOPE OF THE POPULATION REGRESSION LINE   E-743:I-893

A step-by-step method for performing a hypothesis test to decide whether the slope, β_1, of a population regression line is not zero is given in Procedure 15.1 beginning on page 891 of *IS/5e* (Procedure 14.1 on page 741 of *ES/4e*). Alternatively, we can use Excel to carry out such a hypothesis test. In fact, Excel's regression output contains all the information that we need. To illustrate, we return to the Orion example.

EXAMPLE 15.3 *Illustrates the t-test of utility of regression*

In Section 14.2 of this manual we applied the **Regression** data analysis tool to the data on age and price for a sample of 11 Orions shown in Table 14.1 on page 150. Presuming that the variables, age and price satisfy the assumptions for regression inferences (page 158), do the data provide sufficient evidence to conclude that age is useful as a predictor of price for Orions? Perform the test

$$H_0: \beta_1 = 0 \text{ (age is not useful for predicting price)}$$
$$H_a: \beta_1 \neq 0 \text{ (age is useful for predicting price)}$$

at the 5% significance level using the Excel output.

SOLUTION The information we seek here is captured in the cell range **A17:E19** and is duplicated for you in Figure 15.4.

FIGURE 15.4
Excel partial regression output

	Coefficients	Standard Error	t Stat	P-value
Intercept	195.47	15.24	12.83	0.00
AGE	-20.26	2.80	-7.24	0.00

Look at the third line of the computer output in Figure 15.4, the line labeled **AGE**. The second entry in that line, which is under the column headed *Coefficients* gives the slope, b_1, of the sample regression line; hence, $b_1 = -20.261$. The third entry in that line, which is under the column headed *Standard Error* shows the estimated standard deviation of b_1, $s_e/\sqrt{S_{xx}}$. So, $s_e/\sqrt{S_{xx}} = 2.800$.

The fourth entry in that line, which is under the column headed *t Stat*, displays the value of the test statistic

$$t = \frac{b_1}{s_e/\sqrt{S_{xx}}}$$

Thus, we see that $t = -7.237$.

The final entry of the row labeled **AGE**, under the column headed *P-value* gives the P-value for the hypothesis test. This is really the only quantity that we need in order to decide whether the null hypothesis should be rejected. Recall that the P-value is equal to the smallest significance level at which the null hypothesis can be rejected. Therefore, if the P-value is less than or equal to the specified significance level, then we reject H_0; otherwise, we do not reject H_0.

From Figure 15.4, we see that the P-value is 0.000 (to three decimal places). Since this is less than the specified significance level of $\alpha = 0.05$, we reject H_0. In other words, the data provide sufficient evidence to conclude that the slope of the population regression line is not zero and, hence, that age is useful as a predictor of price for Orions. ■

PROBLEM 15.3 In Problem 14.2 on page 152, you performed a regression of weight on height for a sample of males 18–24 years old. Use the output you obtained to decide whether the slope, β_1, of the population regression line is not zero. Perform the hypothesis test at the 5% significance level using Excel. ◊

CONFIDENCE INTERVALS FOR THE SLOPE OF THE POPULATION REGRESSION LINE

E-745 : I-895

We can also employ Excel to help us find a *confidence interval* for the slope, β_1, of the population regression line following the two steps given in Procedure 15.2 on page 894 of *IS/5e* (Procedure 14.2 on page 744 of *ES/4e*). Again, we will use the age and price data for Orions to illustrate the steps, repeated here.

EXAMPLE 15.4 *Illustrates how to use Excel in conjunction with Procedure 15.2 in IS/5e (Procedure 14.2 in ES/4e)*

Use the sample data in Table 14.1 on page 150 to find a 95% confidence interval for the slope, β_1, of the population regression line relating age, x, and price, y, of Orions.

SOLUTION The information we seek here is captured in the cell range **F17:G19** and is duplicated for you in Figure 15.5.

FIGURE 15.5
Excel confidence limits for the slope β_1

Lower 95%	Upper 95%
160.99	229.94
-26.59	-13.93

We can be 95% confident that β_1 is between -26.59 and -13.93. In other words, we can be 95% confident that the yearly drop in mean price for Orions is somewhere between $1393 and $2659.

For a different confidence level, you may add the following step to the menu commands of Example 15.2.

Excel commands:

1 Select the **Confidence Level** check box and type 90 in the text box

The output will then display both the (default) 95% and the 90% confidence interval as shown in Figure 15.6.

FIGURE 15.6
Excel 95% and 90% confidence intervals for β_1

Lower 95%	Upper 95%	Lower 90.0%	Upper 90.0%
160.992	229.945	167.531	223.406
-26.594	-13.928	-25.393	-15.129

You see that the 90% confidence interval now runs from -25.3934 to -15.129. ■

PROBLEM 15.4 Refer to the output of the **Regression** tool that you obtained in Problem 14.2 on page 152. Find a 95% confidence interval for the slope, β_1, of the population regression line relating height, x, and weight, y, of males 18–24 years old. ◊

15.3 Estimation and prediction

In this section, we will learn how to use Excel to help make two important types of inferences. One is to estimate the *conditional mean* of the population of the response variable corresponding to a *particular value* of the predictor variable. The other is to predict the value of the response variable for a particular value of the predictor variable.

We will employ the Orion example yet again to illustrate the pertinent ideas. Recall from the data in Table 14.1 that ages are in years; prices are in hundreds of dollars, rounded to the nearest hundred. These data are entered and appropriately named in the **DATA** workspace as in the previous lesson.

Also recall that the sample regression line for the data in Table 14.1 is determined by the equation $\hat{y} = 195.47 - 20.261x$, using the values given under **Coef**. Thus, our estimate for the mean price of all three-year-old Orions is

$$\hat{y} = 195.47 - 20.261 \cdot 3 = 134.69$$

or $13,469. [Note that the estimate for the mean price of all three-year-old Orions is the same as the predicted price of a randomly selected three-year-old Orion. Both are obtained by substituting $x = 3$ into the sample regression equation.]

The estimate of $13,469 for the mean price of all three-year-old Orions is a point estimate. As we know, it would be more informative if we had some idea of the accuracy of that point estimate. In other words, it would be better to provide a confidence-interval estimate for the mean price of all three-year-old Orions. We will see how Excel can be used to find this confidence interval.

CONFIDENCE INTERVALS FOR CONDITIONAL MEANS

E-753 : I-903

Excel has regression options that may be used indirectly for estimation and prediction. In the **REGRESS** worksheet we have provided a special panel in the range **M2:R20**, entitled **Estimating Means and Predicting Values**, to allow you to compute the required confidence interval. You must name the data in the **DATA** worksheet in order to implement the procedure. And, of course, you must have regression output for that data starting in cell **A2** as we have.

EXAMPLE 15.5 *Illustrates confidence interval estimation for conditional means*

Data on age and price for a sample of 11 Orions are found in Table 14.1 on page 150. Apply Excel to that data to obtain a 95% confidence interval for the mean price of all three-year-old Orions.

SOLUTION   The data are located in columns named AGE and PRICE in the **DATA** worksheet. Open the **REGRESS** worksheet in the Regression workbook. Now execute the Excel commands given on the following page.

Excel commands:

1. Select input cell **N2** and type .05
2. Select input cell **M5** and type 3

With these simple inputs, the output may be found immediately in the range **M3:P5** and should appear as in Figure 15.7.

FIGURE 15.7
Excel output for estimating conditional means

Confidence	95.0%	Confidence	Interval
AGE	Fit	Lower Limit	Upper Limit
3.00	134.685	117.929	151.440

The item headed Fit, gives the point estimate, $\hat{y}_p = 134.685$ ($13,468), for the mean price of all three-year-old Orions or for the price of a randomly selected three-year-old Orion. This is followed by the lower and upper confidence limits for the confidence interval. Hence, a 95% confidence interval for the mean price of all three-year-old Orions is from 117.929 to 151.440. We can be 95% confident that the mean price of all three-year-old Orions is somewhere between $11,793 and $15,144.

Of course you may compute any confidence level by merely changing the input cell for the choice of α. For example, if you were to change that input to 0.10, you may verify that the 90% confidence interval runs from 121.107 to 148.262. We can be 90% confident that the mean price of all three-year-old Orions is somewhere between $12,111 and $14,826. ∎

PROBLEM 15.5 The height and weight data for a sample of males 18–24 years old may be found in the table for Problem 14.1 on page 152. Use Excel to determine a 95% confidence interval for the mean weight of all males 18–24 years old who are 70 inches tall.

PREDICTION INTERVALS E-756 : I-906

One of the main reasons for determining the regression equation for a sample of data points is to use it for making predictions. Procedure 15.4 located on page 905 of *IS/5e* (Procedure 14.4 located on page 755 of *ES/4e*) gives a step-by-step method for obtaining a prediction interval for the value of the response variable corresponding to a particular value of the predictor variable. You may already have noticed that this is part of the output for our **Estimating Means and Predicting Values** panel. Consider the following example.

EXAMPLE 15.6 *Illustrates prediction intervals*

The age and price data for a sample of Orions are given in Table 14.1 on page 150. Apply Excel to that data to obtain a 95% prediction interval for the price of a randomly selected three-year-old Orion.

SOLUTION In Example 15.5 we found a 95% confidence interval for the mean price of all three-year-old Orions. The output of the **Estimating Means and Predicting Values** panel also provides the desired 95% prediction interval in the range **Q3:R5** for this particular value. Thus, a 95% prediction interval for the price of a randomly selected three-year-old Orion is from 101.667 to 167.702. We can be 95% certain that the price of a randomly selected three-year-old Orion will be somewhere between $10,167 and $16,770. Contrast this with the confidence interval for the mean of all such Orions.

All this and more is displayed in Figure 15.8.

FIGURE 15.8
Excel output for Estimating Means and Predicting Values

$\alpha =$	0.05	Estimating Means and Predicting Values			
Confidence	95.0%	Confidence	Interval	Prediction	Interval
AGE	Fit	Lower Limit	Upper Limit	Lower Limit	Upper Limit
3.00	134.685	117.929	151.440	101.667	167.702
3.50	124.554	110.425	138.683	92.789	156.319
4.00	114.423	102.653	126.194	83.634	145.212

As you see from the panel, you may enter other choices for the predictor variable, as many as you please (with caution about the extrapolation problem), under the column headed **AGE**. The panel will display all confidence intervals and prediction intervals for those choices instantly. ■

(Note: Remember to name your data sets for re-use of this panel.*)*

PROBLEM 15.6 Refer to Problem 15.5 on page 166. Use the output you obtained in Problem 15.5 to find a 95% prediction interval for the weight of a randomly selected male 18–24 years old who is 70 inches tall. ◇

15.4 Inferences in correlation `E-765:I-915`

In the previous section we learned that we can decide whether two variables, x and y, are linearly related by performing a hypothesis test for the slope, β_1, of the population regression line. Alternatively, we can perform a hypothesis test for the *population linear correlation coefficient*, ρ (rho).

The population linear correlation coefficient, ρ, measures the linear correlation between the population of *all* data points in the same way that the (sample) linear correlation coefficient, r, measures the linear correlation between a sample of data points. Thus, it is ρ that actually describes the linear correlation between two variables.

A step-by-step method for performing a hypothesis test with null hypothesis $H_0 : \rho = 0$ (that is, the variables are linearly uncorrelated) is given by Procedure 15.5 on page 913 of *IS/5e* (Procedure 14.5, page 763 of *ES/4e*). Rejection of the null hypothesis indicates that the variables under consideration are linearly related.

The test statistic for the hypothesis test is a function of the (sample) correlation coefficient, r. We discovered in Section 14.3 of this manual that Excel can be used to compute the value of r so that it should be a simple matter to construct the corresponding t-test. To allow for the utmost versatility, we have created a panel labeled **Correlation t-test** in the range **T2:W6** of the **REGRESS** worksheet as a program independent of any regression output. All you need to do is input the names of two paired data sets, the desired level of significance for the test, and an Alternative code. The next example illustrates its utility.

EXAMPLE 15.7 *Illustrates how Excel can be used for a correlation test*

Consider once more the age and price data for Orions. Do the data provide sufficient evidence to conclude that the variables age, x, and price, y, for Orions are *negatively* linearly correlated? Use $\alpha = 0.05$.

SOLUTION Open the **REGRESS** worksheet and proceed as follows.

Menu commands:

1 Select input cell **T5** and type `AGE`
2 Select input cell **T6** and type `PRICE`
3 Select input cell **V3** and type `.05`
4 Select input cell **V4** and type `-1`

The output is displayed in Figure 15.9.

FIGURE 15.9
Excel output for Correlation t-test

DATA FILE	Correlation	t-Test	
	$\alpha =$	0.05	
NAMES:	Alt Code	-1	P-value
AGE	Correlation	t-statistic	0.000
PRICE	-0.924	-7.237	Reject

With a *P*-value of 0.000, we surely reject the hypothesis that the variables age and price for Orions are linearly uncorrelated, electing to conclude that they are negatively linearly correlated. In other words, the price of a Orion tends to decrease linearly as its age increases, at least for Orions between two and seven years old where our data are located. ∎

It is worthwhile to note that the output in cell **V6** and the output in cell **D19** in the regression output are identical. This is no mere coincidence; they are mathematically equivalent. That is to say, in a regression model, the *t*-test for correlation is redundant since testing for a slope of 0 leads to exactly the same conclusion. The Correlation procedure is more general, however, since not all sets of paired data satisfy the requirements for a regression analysis.

PROBLEM 15.7 Consider again the height and weight data for a sample of males 18–24 years old treated in Problem 15.4 on page 164. Do the data provide sufficient evidence to conclude that the variables height and weight for males 18–24 years old are *positively* linearly correlated? Perform the test at the 5% significance level using Excel. ◊

LESSON 16

Analysis of Variance (ANOVA)

GENERAL OBJECTIVE In Lesson 10 we saw how Excel can be used to draw inferences concerning the means of two populations. Now we will examine how Excel can be applied to draw inferences concerning the means of more than two populations. The statistical procedures for comparing the means of more than two populations are referred to as *analysis of variance*, or *ANOVA*.

LESSON OUTLINE
16.1 The F-distribution
16.2 One-way analysis of variance

16.1 The F-distribution

E-692:I-936

The analysis-of-variance procedures utilize a class of continuous probability distributions called *F-distributions*, named in honor of Ronald Fisher (1890–1962). We will now study how Excel can be used to find percentile points, F_α, for this distribution. These are required by analysis of variance.

To perform an analysis of variance, we need to be able to find, for a given F-curve, the F-value having a specified area to its right. This can be accomplished using Excel's FINV function in conjunction with a specification of degrees of freedom, a pair of numbers in each case. We explain the details in the following example.

EXAMPLE 16.1 *Illustrates finding F percentiles*

For an F-curve with df = (4, 12), find $F_{0.05}$. That is, find the F-value with area 0.05 to its right.

SOLUTION Select any empty cell in a worksheet and type =FINV(.05,4,12). The value 3.2592 (depending on the number formatting of the cell) should appear instantly. Thus, for an F-curve with df = (4, 12), $F_{0.05} = 3.2592$. ∎

PROBLEM 16.1 For an F-curve with df = (12, 5), find
a) $F_{0.05}$ b) $F_{0.01}$ c) $F_{0.025}$ ◇

16.2 One-way analysis of variance

E-707:I-951

We have already mentioned that analysis of variance (ANOVA) is an inferential procedure that is used to compare the means of several populations. The reason for the word "variance" in "analysis of variance" is that the procedure for comparing the means involves analyzing the variation in the sample data.

In this section we will examine *one-way analysis of variance*. It is called *one-way* analysis of variance because each piece of data is classified in one way; namely, according to the population from which it was sampled. As in the pooled-t procedure, we make the following assumptions:

KEY FACT 16.1 Assumptions for one-way ANOVA

1. *Independent samples:* The samples taken from the populations under consideration are independent of one another.

2. *Normal populations:* The populations under consideration are normally distributed.

3. *Equal standard deviations:* The standard deviations of the populations under consideration are equal.

Lesson 16 Analysis of Variance (ANOVA)

Assumption 1, the independent-samples assumption, is often referred to as the assumption of a *completely randomized design*. Procedure 16.1 located on page 948 of *IS/5e* (Procedure 13.1 located on page 704 of *ES/4e*) provides a step-by-step method for performing a one-way ANOVA test. Alternatively, we can apply Excel to carry out such a hypothesis test using a special procedure in the **Data Analysis Tool**. The next example explains how this option is used.

EXAMPLE 16.2 Illustrates ANOVA in Excel

The U.S. Energy Information Administration gathers data on residential energy consumption and expenditures. A researcher wants to know if there is a difference in mean annual energy consumption among the four regions of the United States. Independent random samples of households in the four U.S. regions yield the data on last year's energy consumption given in Table 15.1. Use Excel to perform the hypothesis test at the 5% significance level.

TABLE 16.1 Samples and their means of last year's energy consumption for households in the four U.S. regions

Northeast	Midwest	South	West
15	17	11	10
10	12	7	12
13	18	9	8
14	13	13	7
13	15		9
	12		
13.0	14.5	10.0	9.2	← Means

SOLUTION Let μ_1, μ_2, μ_3, and μ_4 denote last year's mean energy consumption for households in the Northeast, Midwest, South, and West, respectively. Then the null and alternative hypotheses are

H_0: $\mu_1 = \mu_2 = \mu_3 = \mu_4$ (mean energy consumption are equal)

H_a: Not all the means are equal.

To employ Excel, we have entered the four data sets of Table 16.1 into columns named **NRTHEAST**, **MIDWEST**, **SOUTH** and **WEST** in the **DATA** worksheet of the **Regression** workbook. Open the **ANOVA** worksheet in that workbook and proceed as follows.

Excel commands:

1. Choose **Tools ▶ Data Analysis...**
2. Select **Anova:Single Factor** from the **Analysis Tools**
3. Use the button in the **Input Range:** text box to select the range **E1:H7** in the **DATA** worksheet
4. Select the **Columns** option button for the **Grouped by:** options
5. Select the **Labels in First Row** check box and type **.05** in the **Alpha:** text box
6. Click the **Output Range:** radio button in Output options

7 Use the ▦ button in the **Output Range:** text box to select cell **A1** in the **ANOVA** worksheet

8 Click [OK]

Excel's version of the one-way ANOVA table in Table 16.5 on page 946 of *IS/5e* (Table 13.5 on page 702 of *ES/4e*) may be found in the range **A11:G16** and is reproduced here in Figure 16.1 (with some cell formatting of our own). Note that Excel uses the terminology "Between Groups" instead of "Treatment" and "Within Groups" instead of "Error."

FIGURE 16.1
Excel (partial) ANOVA output

ANOVA

Source of Variation	SS	df	MS	F	P-value	F crit
Between Groups	97.500	3	32.500	6.318	0.005	3.239
Within Groups	82.300	16	5.144			
Total	179.800	19				

From the figure, we see that the P-value is 0.005, which is less than the specified significance level of $\alpha = 0.05$, so we reject H_0. In other words, the data provide sufficient evidence to conclude that last year's mean energy consumption for households in the four U.S. regions are not all the same.

We may also use Figure 16.1 to carry out the various steps of Procedure 16.1 on page 948 of *IS/5e* (Procedure 13.1 on page 704 of *ES/4e*). The two numbers of degrees of freedom for the appropriate F-curve are displayed in the column headed **df**. Thus, df = (3, 16). We are to perform the hypothesis test at the 5% significance level; so $\alpha = 0.05$. Consequently, the critical value is $F_{0.05}$ for an F-curve with df = (3, 16). This value is listed under the heading Fcrit and is equal to 3.239. The value of F is displayed in the column headed F in Figure 16.1. Thus, $F = 6.318$. Since the value of the test statistic, $F = 6.318$, exceeds the critical value, $F_{0.05} = 3.239$, we reject H_0 as before. ■

OTHER ITEMS IN THE OUTPUT

We will now examine some of the other items in the output shown in Figure 16.1. Recall the fundamental identity in one-way analysis of variance is the following *one-way ANOVA identity:*

$$SST = SSTR + SSE$$

This identity shows that, indeed, the total variation among all the sample data can be partitioned into a component representing variation among sample means and a component representing variation within samples. The Excel (partial) output in Figure 16.1 displays the values of the three sums of squares—*SST*, *SSTR*, and *SSE*.

To illustrate, examine Figure 16.1 once more. The three sums of squares are printed in the third column, which is headed SS (for sums of squares). Hence, we see that for the energy-consumption data, $SSTR = 97.50$ (Between Groups), $SSE = 82.30$ (Within Groups), and $SST = 179.80$ (Total). Note that for these sums of squares

$$179.80 = 97.50 + 82.30$$

This illustrates the one-way ANOVA identity.

Above the ANOVA table (Figure 16.1) is a SUMMARY table that gives the sample sizes, sample means, and sample variance of the four samples. This table is displayed in Figure 16.2.

FIGURE 16.2
Excel (partial) ANOVA output

SUMMARY				
Groups	Count	Sum	Average	Variance
NRTHEAST	5	65.000	13.000	3.500
MIDWEST	6	87.000	14.500	6.700
SOUTH	4	40.000	10.000	6.667
WEST	5	46.000	9.200	3.700

With the average given for each of the groups, we may proceed to compute confidence intervals for the corresponding population group means.

EXAMPLE 16.3 *Illustrates ANOVA confidence intervals in Excel*

For the data on residential energy consumption and expenditures of Example 16.2, find 95% confidence intervals for the mean of each region included in the study.

SOLUTION   The pooled estimate of the common population standard deviation, σ, of the four populations is just the square root of the mean square (MS) within groups. The latter is found in cell **D14** of the output and is equal to 5.144 (see Figure 16.1). This is the pooled estimate, s, of the common population standard deviation, σ, of the four populations. This quantity is needed to compute confidence intervals as discussed right after Exercise 16.42 of *IS/5e* (Exercise 13.42 of *ES/4e*).

We have created a special panel in the **ANOVA** worksheet under columns **I, J** and **K** to compute the confidence interval for each mean under study. This panel has three input cells, denoted as usual by dotted borders. Proceed as follows.

Excel commands:

1 Type <u>16</u> (the Within Groups df found in cell **C14**) in input cell **I3**
2 Type <u>5.144</u> (the Within Groups MS found in cell **D14**) in input cell **J3**
3 Type <u>.05</u> in input cell **K3**
4 Select the range **I6:K7** and drag the fill handle of **K7** through as many rows as there are group means (row 9 in this example)

The confidence intervals may then be read directly off the table shown in Figure 16.3 on the following page for each one of the group means.

FIGURE 16.3
Excel confidence interval output

Within df	MS	α
16	5.144	0.05

95.0% Confidence Intervals		
Groups	Lower Limit	Upper Limit
NRTHEAST	10.850	15.150
MIDWEST	12.537	16.463
SOUTH	7.596	12.404
WEST	7.050	11.350

Thus, for example, with 95% confidence we may state that the mean energy consumption for the South is somewhere between 7.596 and 12.404. ∎

PROBLEM 16.2 A chain of convenience stores wanted to test three different advertising policies:

Policy 1: No advertising.
Policy 2: Advertise in neighborhoods with circulars.
Policy 3: Use circulars and advertise in newspapers.

Eighteen stores were randomly selected and divided randomly into three groups of six stores. Each group used one of the three policies. Following the implementation of the policies, sales figures were obtained for each of the stores during a 1-month period. The figures are displayed in the following table in thousands of dollars.

Policy 1	Policy 2	Policy 3
22	21	29
20	25	24
26	25	31
21	20	32
24	22	26
22	26	27

Do the data provide sufficient evidence to conclude that mean sales for the three advertising policies are not all the same? They are stored in the **DATA** worksheet of the **Regression** workbook under their names given here. Use $\alpha = 0.05$. Apply Excel to perform the appropriate hypothesis test. Use the output you obtain to find the three sums of squares, $SSTR$, SSE, and SST, for the sales data. Also determine the pooled estimate of the common standard deviation from the Excel output and find 95% confidence intervals for each Policy mean. ◇

APPENDIX A

Answers to Problems

In this appendix you will find the answers to all of the problems given in this manual. The answers are grouped by Lessons according to the problem numbers given in the lesson. This should make it easy for you to find the problem you are interested in. Most of the time the answers will involve Excel printouts.

Lesson 1

PROBLEM 1.1 Answers will vary.

PROBLEM 1.2 Answers will vary.

Lesson 2

PROBLEM 2.1 Excel output:

Bin	Frequency	Rel. Freq.	Midpoint
40.0	3	0.075	35.0
50.0	1	0.025	45.0
60.0	8	0.200	55.0
70.0	10	0.250	65.0
80.0	7	0.175	75.0
90.0	7	0.175	85.0
100.0	4	0.100	95.0

PROBLEM 2.2 Excel output:

Bin	Frequency	Rel. Freq.	Midpoint
0	0	0.000	0.0
1	7	0.175	1.0
2	13	0.325	2.0
3	9	0.225	3.0
4	5	0.125	4.0
5	4	0.100	5.0
6	1	0.025	6.0
7	1	0.025	7.0

PROBLEM 2.3 Excel output:

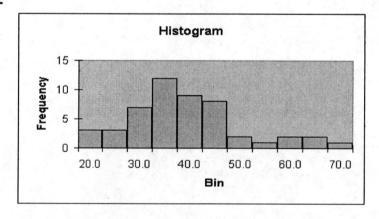

Hint: Format the Category Axis using the Scale tab to display two categories between tick marks.

PROBLEM 2.4   Excel output:

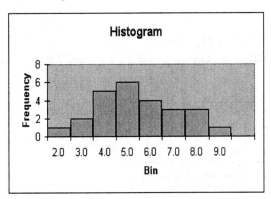

PROBLEM 2.5   Excel output:

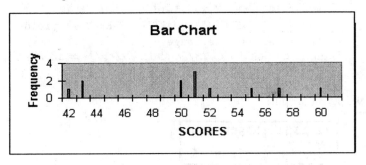

Lesson 3

PROBLEM 3.1   Excel output:
```
=AVERAGE(LT2) yields      63.5
=AVERAGE(GTH2) yields     74.8
```

PROBLEM 3.2   Excel output:
```
=MEDIAN(LT2) yields       62.5
=MEDIAN(GTH2) yields      75.0
```

PROBLEM 3.3   Excel output:

```
=MODE(PATIENTS) yields    5
```
The mode of the number of rooms is 5.

A-4 Answers to problems

PROBLEM 3.4 Excel output:

=SUM(PATIENTS) yields   229.3

Hence the total number of patients admitted to U.S. hospitals between 1977 and 1982 was 229.3 million.

PROBLEM 3.5 Excel output:

=SUM(FBI)/COUNT(FBI) yields   78.5
=AVERAGE(FBI) yields   78.5

PROBLEM 3.6 Excel output:

=MAX(FBI)-MIN(FBI) yields   141

The range is $141 in theft losses.

PROBLEM 3.7 Excel output:

=STDEV(FBI) yields   39.212

PROBLEM 3.8 Excel output:

=QUARTILE(ENERGY,3) yields   1382.50
=QUARTILE(ENERGY,1) yields   1107.25
=QUARTILE(ENERGY,3)-QUARTILE(ENERGY,1) yields 275.25,
  the interquartile range.

Using Definition 4.7 of *IS/5e* and *ES/4e*, $Q_3 = 1389.5$, $Q_1 = 1103.75$ and hence $IQR = 285.75$

PROBLEM 3.9 Excel output:

DATA FILE: ENERGY	DESCRIPTIVE MEASURES
n =	36
Mean =	1243.639
Stdev =	165.135
SE Mean =	27.522
Min =	949.000
Q_1 =	1103.75
Median =	1231.000
Q_3 =	1389.5
Max =	1532.000
IQR =	285.750

PROBLEM 3.10 Excel output:

=AVERAGE(HTS) yields   74.1

The population mean height is 74.1 inches.

Lesson 4

PROBLEM 3.11 Excel output:

=STDEVP(HTS) yields   2.93

The population standard deviation is 2.93 inches.

Lesson 4

PROBLEM 4.1 Using the output in Figure 4.1 (page 45):

a) Excel output:

=C2+C3 yields 0.140

b) Excel output:

=C3+C4+C5 yields 0.350

c) Excel output:

=C2+C8 yields 0.261

PROBLEM 4.2 Excel output:

Sum of NUMBER	PARTY		
YEARS	Democrat	Republican	Grand Total
Under 2	59	25	84
2--9	134	93	227
10--19	44	41	85
20--29	20	8	28
30&over	10	0	10
Grand Total	267	167	434

PROBLEM 4.3 Excel output:

Joint Probability	PARTY		
YEARS	Democrat	Republican	Grand Total
Under 2	0.136	0.058	0.194
2--9	0.309	0.214	0.523
10--19	0.101	0.094	0.196
20--29	0.046	0.018	0.065
30&over	0.023	0.000	0.023
Grand Total	0.615	0.385	1.000

The probability of selecting a congressman age 20–29 is 0.065 according to the table.

PROBLEM 4.4 Excel output:

Conditioned on YEARS	PARTY			Conditioned on PARTY	PARTY		
YEARS	Democrat	Republican	Grand Total	YEARS	Democrat	Republican	Grand Total
Under 2	0.702	0.298	1.000	Under 2	0.221	0.150	0.194
2–9	0.590	0.410	1.000	2–9	0.502	0.557	0.523
10–19	0.518	0.482	1.000	10–19	0.165	0.246	0.196
20–29	0.714	0.286	1.000	20–29	0.075	0.048	0.065
30&over	1.000	0.000	1.000	30&over	0.037	0.000	0.023
Grand Total	0.615	0.385	1.000	Grand Total	1.000	1.000	1.000

The probability of a Republican among those in office for 30 or more years is 0; the probability of a congressman being in office under 2 years among Democrats is 0.221.

Lesson 5

PROBLEM 5.1 Excel output:

y	f	P(Y=y)
1	19.4	0.232
2	26.5	0.317
3	14.6	0.175
4	12.9	0.154
5	6.1	0.073
6	2.5	0.030
7	1.6	0.019

PROBLEM 5.2 Answers will vary.

PROBLEM 5.3 Using the output in Problem 5.1, copy the data from y to x and from P(Y=y) to P(X=x) in the **DATA** worksheet. From cell **O2** read the mean value of $\mu_x = 2.685$ and from cell **Q2** read the standard deviation value of $\sigma_x = 1.472$.

PROBLEM 5.4 a) Excel output:

=BINOMDIST(2,3,0.25,FALSE) yields 0.141

b) Excel output:

=BINOMDIST(2,3,0.25,TRUE) yields 0.984

c) Excel output:

=1-BINOMDIST(0,3,0.25,TRUE) yields 0.578

d), e) Excel output:

x	P(X=x)
0	0.422
1	0.422
2	0.141
3	0.016

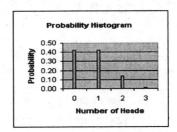

Skewness is markedly to the right. As a bonus, observe in cells **O2** and **P2** that both the mean and standard deviation are equal to 0.750 to three decimal places.

PROBLEM 5.5 a) Excel output:

=POISSON(2,1,FALSE) yields 0.184

b) Excel output:

=POISSON(2,1,TRUE) yields 0.920

c), d) Excel output:

x	P(X=x)
0	0.368
1	0.368
2	0.184
3	0.061
4	0.015
5	0.003
6	0.001
7	0.000

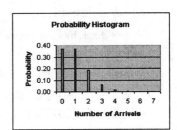

Skewness is markedly to the right. As a bonus, observe in cells **O2** and **P2** that both the mean and standard deviation are both equal to 1.

PROBLEM 5.6 a) Excel output:

=POISSON(2,225,FALSE) yields 0.2668
=BINOMDIST(2,500,0.0045,FALSE) yields 0.2673

b) Excel output:

=POISSON(4,2.25,TRUE) yields 0.9220
=BINOMDIST(4,500,0.0045,TRUE) yields 0.9224

Lesson 6

PROBLEM 6.1 Answers will vary.

PROBLEM 6.2 a) Excel output:

=NORMSDIST(-1.28) yields 0.1003

b) Excel output:

=NORMSDIST(2.47)-NORMSDIST(1.53) yields 0.0563

c) Excel output:

=1-NORMSDIST(1.64) yields 0.0505

d) Excel output:

=NORMSDIST(-1.85)+1-NORMSDIST(.29) yields 0.4181

PROBLEM 6.3 a) Excel output:

=NORMSINV(.75) yields 0.6745

b) Excel output:

=NORMSINV(.05) yields -1.6449

c) Excel output:

=NORMSINV(.95) yields 1.6449

Combining this with b), there is a middle area of 90% between $z = -1.6449$ and $z = 1.6449$. Note that $z_{0.95} = -z_{0.05}$.

d) Excel output:

=NORMSINV(.96) yields 1.5548

So, $z_{0.06}=1.5548$ and, as in c), $z_{0.94} = -1.5548$.

PROBLEM 6.4 a) Excel output:

=NORMDIST(12,15,2,TRUE) yields 0.0668

b) Excel output:

=NORMSDIST(18,15,2,TRUE)-NORMSDIST(14,15,2,TRUE) yields 0.6247

c) Excel output:

=1-NORMSDIST(16.5,15,2,TRUE) yields 0.2266

PROBLEM 6.5 a) Excel output:

=NORMINV(.90,64.6,2.4) yields 67.68

b) Excel output:

=NORMSINV(.10,64.6,2.4) yields 61.52

PROBLEM 6.6 Excel output:

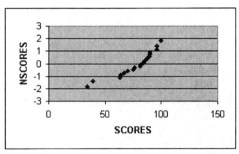

The departure from linearity is not too severe.

PROBLEM 6.7 Excel output:

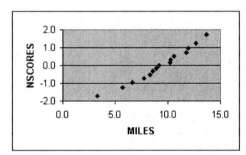

The departure from linearity is not too severe.

Lesson 7

PROBLEM 7.1 Answers will vary.

PROBLEM 7.2 Answers will vary.

PROBLEM 7.3 Answers will vary.

Lesson 8

PROBLEM 8.1 Excel output from **One Mean** panel in **1SAMP** worksheet:

z-Confidence	Interval:
90%	**Two-sided:**
Lower Limit l =	133.546
Upper Limit u =	140.204

We can be 90% confident that the mean weight, μ, of all U.S. women 5 feet 4 inches tall and in the age group 18–24 is somewhere between 133.55 and 140.20 pounds.

PROBLEM 8.2 a) Excel output:

```
=TINV(0.2,8) yields 1.3968
```

b) Excel output:

```
=TINV(0.01,8) yields 3.3554
```

c) Excel output:

```
=TINV(0.02,8) yields 2.8965
```
which is $t_{.01}$. Hence, $t_{0.99} = -2.8965$

PROBLEM 8.3 Excel output from **One Mean** panel in **1SAMP** worksheet:

t-Confidence	Interval:
95%	**Two-sided:**
Lower Limit l =	117.385
Upper Limit u =	128.115

We can be 95% confident that the mean weight, μ, of this year's mean annual subscription rate is somewhere between $117.39 and $128.12.

Lesson 9

PROBLEM 9.1 Excel output from **One Mean** panel in **1SAMP** worksheet:

One Mean			
DATA FILE NAME:		$\sigma =$ 11.600	
TRAVTIME	n = 35	df =	
	xbar = 14.029	Min =	0.00
	s = 13.321	$Q_1 =$	4.5
	SE Mean 1.96	Median =	10
	$H_0: \mu_0 =$ 13.00	$Q_2 =$	20
	Alt Code = 0	Max =	52.00
z-Test:	$\alpha =$ 0.05	**z-Confidence**	**Interval:**
Two-tailed test: z = 0.525		95%	Two-sided:
Lower Critical Value: -1.960		Lower Limit l =	10.186
Upper Critical Value: 1.960		Upper Limit u =	17.872
P-value: 0.5999		$\mu_0 =$	13.00
Action: Do not Reject		Action:	Do not Reject

The data do not provide sufficient evidence to conclude that this year's mean travel time to work for South Dakota residents have changed from the 1980 mean of 13 minutes.

PROBLEM 9.2 Excel output from **One Mean** panel in **1SAMP** worksheet:

One Mean			
DATA FILE NAME:		$\sigma =$	
SAT	n = 25	df =	24
	xbar = 437.440	Min =	287.00
	s = 86.676	$Q_1 =$	376
	SE Mean 17.34	Median =	434
	$H_0: \mu_0 =$ 428.00	$Q_2 =$	494
	Alt Code = 1	Max =	613.00
t-Test	$\alpha =$ 0.1	**t-Confidence**	**Interval:**
Right-tailed test: t = 0.545		90%	One-sided:
Critical Value: 1.318		Lower Limit l =	414.595
P-value: 0.2955		$\mu_0 =$	428.00
Action: Do not Reject		Action:	Do Not Reject

The data do not provide sufficient evidence to conclude that last year's mean for verbal SAT scores has increased over the 1983 mean of 430 points.

Lesson 10

PROBLEM 10.1 Excel output from **Two Means** panel in **2SAMP** worksheet:

	Two Means			
DATA FILE NAMES:	n_1 =	10	n_2 =	10
NEW	$\bar{x}_1$ =	42.130	$\bar{x}_2$ =	43.230
PRESENT	s_1 =	0.685	s_2 =	0.750
	sp =	0.718	df =	18
	H_0: μ_d =	0.00	σ's Equal ?	y
Pooled	Alt Code =	-1	α =	0.05
Left-tailed t-test: t =	-3.425		**95%**	**One-sided:**
Critical Value:	-1.734			
			Upper Limit u =	-0.543
P-value:	0.0015		μ_d =	0
Action:	Reject		**Action:**	Reject

Evidently, the new machine packs faster, on the average since the hypothesis that there is no difference is rejected.

PROBLEM 10.2 Excel output from **Two Means** panel in **2SAMP** worksheet:

	Two Means			
DATA FILE NAMES:	n_1 =	15	n_2 =	15
SURE	$\bar{x}_1$ =	89.933	$\bar{x}_2$ =	92.267
MIRROR	s_1 =	1.944	s_2 =	1.033
	s =	0.568	df =	21
	H_0: μ_d =	0.00	σ's Equal ?	n
Nonpooled	Alt Code =	-1	α =	0.01
Left-tailed t-test: t =	-4.104		**99%**	**One-sided:**
Critical Value:	-2.518			
			Upper Limit u =	-0.902
P-value:	0.0003		μ_d =	0
Action:	Reject		**Action:**	Reject

The data indicate that Mirror-Sheen has a longer effectiveness time, on the average, than Sureglow.

PROBLEM 10.3 Excel output from **One Mean** panel in **1SAMP** worksheet:

DATA FILE NAME:		**One Mean**		
		σ =		
DIFFER	n =	10	df =	9
	xbar =	-0.900	Min =	-4.00
	s =	2.514	Q_1 =	-2
	SE Mean	0.80	Median =	-1.5
	$H_0: \mu_0$ =	0.00	Q_2 =	-0.25
	Alt Code =	-1	Max =	5.00
t-Test	α =	0.05	t-Confidence	Interval:
Left-tailed test: t =	-1.132		95%	One-sided:
Critical Value:	-1.833			
			Upper Limit u =	0.558
P-value:	0.1435		μ_0 =	0.00
Action:	Do not Reject		Action:	Do Not Reject

The data do not provide sufficient evidence to conclude that the new variety of wheat gives a better yield, on the average, than the current variety.

Lesson 11

PROBLEM 11.1
a) Excel output:
=CHIINV(0.1,14) yields 21.064
b) Excel output:
=CHIINV(0.9,8) yields 7.7895

PROBLEM 11.2
a) Excel output:
=CHIDIST(13,14) yields 0.5625
b) Excel output:
=CHIDIST(7,8)-CHIDIST(13,14) yields 0.4082

PROBLEM 11.3 Consult the output for the **One Standard Deviation** panel on the following page. On the basis of the data, we cannot conclude that the standard deviation of this year's verbal scores is different from the 1941 standard deviation of 100.

PROBLEM 11.4 Consult the output for the **One Standard Deviation** panel on the following page. You see that a 95% confidence interval for the standard deviation, σ, of this year's verbal scores runs from 86.658 to 154.393.

PROBLEM 11.5
a) Excel output:
=FINV(.05,12,5) yields 4.6777
b) Excel output:
=FINV(.01,12,5) yields 9.8883
c) Excel output:
=FINV(.025,12,5) yields 6.5245
d) Excel output:
=FDIST(1.5,12,5) yields 0.3440

Excel output from **One Standard Deviation** panel in **1SAMP** worksheet:

	One Standard Deviation			
DATA FILE NAME:	n =	25	Min =	228.00
VERBAL	s =	110.9824	Q_1 =	370
	df =	24	Median =	432
	$H_0: \sigma_0$ =	100.00	Q_2 =	558
	Alt Code =	0	Max =	648.00
	α =	0.05		
χ^2-Test Stat :	29.561		χ^2-Confidence	Interval:
Lower Critical Value:	12.401		Lower Limit l =	86.658
Upper Critical Value:	39.364		Upper Limit u =	154.393
P-value:	0.3995		σ_0 =	100.00
Action:	Do not Reject		Action:	Do not Reject

PROBLEM 11.6   Excel output from **Two Standard Deviations** panel in **2SAMP** worksheet:

	Two Standard Deviations			
DATA FILE NAMES:	n_1 =	12	n_2 =	12
S	s_1^2 =	0.345	s_2^2 =	1.813
U	s_1 =	0.587	s_2 =	1.346
	df_1 =	11	df_2 =	11
	$H_0: \sigma_1/\sigma_2$ =	1.00	s_1/s_2 =	0.4360
	α =	0.05	Alt Code =	-1
F-Test Stat :	0.190		F-Conf. Interval	for σ_1/σ_2
Critical Value:	0.355			
			Upper Limit u =	0.732
P-value:	0.0052		σ_1/σ_2 =	1.00
Action:	Reject		Action:	Reject

From the output you see that there is evidence that rainfall is less variable in regions that were seeded.

PROBLEM 11.7   From the above output, an upper confidence bound on the ratio σ_1/σ_2 is 0.732. Another way to put is that we have 95% confidence that $\sigma_1 < 0.7\sigma_2$.

Lesson 12

PROBLEM 12.1 Excel output from **One Prop** panel in **1SAMP** worksheet:

z-Confidence	Interval:
99.0%	Two-sided:
Lower Limit l =	0.784
Upper Limit u =	0.836

We can be 99% confident that the population proportion in favor of voluntary withholding of life support systems from the terminally ill is somewhere between 0.784 and 0.836.

PROBLEM 12.2 Excel output from **One Prop** panel in **1SAMP** worksheet:

Exact-Confidence	Interval:
95.0%	Two-sided:
Lower Limit l =	0.218
Upper Limit u =	0.660

We can be 95% confident that the population proportion of white offspring lies somewhere between 0.218 and 0.660 and this does include 0.25.

PROBLEM 12.3 Excel output from **One Prop** panel in **1SAMP** worksheet:

z-Test		SE test :	0.024	z-Confidence	Interval:
Right-tailed test: z =	2.449			99.0%	One-sided:
Critical Value:	2.326			Lower Limit l =	0.090
P-value:	0.0072			p_0 =	0.10
Action:	Reject			Action:	Do Not Reject

The null hypothesis should be rejected at the 1% level because the P-value is only 0.007. Yet you would not reject the null hypothesis based on confidence intervals since 0.1 is greater than the lower one-sided 99% confidence bound of 0.09. This is one of those cases where the two procedures are not quite equivalent.

However, the exact test procedure yields the following output:

Exact Test			Exact-Confidence	Interval:
Right-tailed test: x =	24.000		99.0%	One-sided:
			Lower Limit l =	0.097
P-value:	0.0143		p_0 =	0.10
Action:	Do not Reject		Action:	Do Not Reject

With a more exact computation of the P-value, you would not reject by either criteria this time and would conclude that this year's percentage of active licensees is not higher.

A-16 Answers to problems

PROBLEM 12.4 Excel output from **Two Proportions** panel in **2SAMP** worksheet:

Two Proportions				
DATA FILE NAMES:	n_1 = 300		n_2 = 250	
SD	x_1 = 184		x_2 = 148	
SD2	$p_1$hat = 0.613		$p_2$hat = 0.592	
	phat = 0.604		SE = 0.0419	
	α = 0.05		Alt Code = 0	
z-Test			z-Confidence Interval:	
Two-tailed test: z = 0.509			95%	Two-sided:
Lower Critical Value: -1.960			Lower Limit l = -0.061	
Upper Critical Value: 1.960			Upper Limit u = 0.103	
P-value: 0.6105			$p_1 - p_2$ = 0.000	
Action: Do not Reject			Action: Do not Reject	

The evidence is not sufficient to conclude that there was a real difference. We can be 95% confident that the difference, $p_1 - p_2$, is somewhere between -0.061 and 0.103 and this includes 0.

Lesson 13

PROBLEM 13.1 See the solution to Problem 11.1, page A-13.

PROBLEM 13.2 See the solution to Problem 11.2, page A-13.

PROBLEM 13.3 Excel output from **GOF** panel in **CHISQ** worksheet:

DATA:				GOF	
Rel Freq	O	E	$(O-E)^2/E$	df = 5	
0.567	91	141.75	18.16975	α = 0.05	
0.143	16	35.75	10.91084	χ^2 = 153.645	
0.16	110	40	122.5	χ^2_α = 11.070	
0.045	14	11.25	0.672222	P-value: 0.000	
0.067	17	16.75	0.003731	Action:	Reject
0.018	2	4.5	1.388889		

Assumptions (1) and (2) are satisfied. For all of the expected frequencies are at least 1. Also, only 1 out of 6 or 16.7% of the expected frequencies are less than 5. It appears that the primary-heating-fuel distribution for occupied housing units built after 1974 differs from that of all occupied housing units.

PROBLEM 13.4 Produces the same output.

PROBLEM 13.5 Answers will vary.

PROBLEM 13.6 Excel output from χ^2 Test for Independence panel in **CHISQ** worksheet:

χ^2 Test for Independence	
$\alpha =$	0.05
df =	2
P-value =	0.539
$\chi^2_\alpha =$	5.991
$\chi^2 =$	1.235
Action	Do not reject

With a P-value of 0.539 the data do not provide sufficient evidence to conclude that the sex of a respondent and their reply on the issue of approval are dependent.

Lesson 14

PROBLEM 14.1 Excel output from the **REGRESS** worksheet:

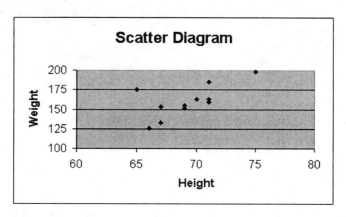

PROBLEM 14.2 Excel output from the **REGRESS** worksheet:

	Coefficients
Intercept	-174.493
HEIGHT	4.836

The required regression equation: PRICE = -174 + 4.836 HEIGHT

PROBLEM 14.3 From the Regression Statistics panel shown on the following page opposite R Square, we have $r^2 = 0.439$ or 43.9%.

Excel output for Regression Statistics

Regression Statistics	
Multiple R	0.662
R Square	0.439
Adjusted R Square	0.376
Standard Error	16.477
Observations	11

PROBLEM 14.4 Excel output from the ANOVA panel:

ANOVA

	df	SS	MS	F	Significance F
Regression	1	1909.460	1909.460	7.033	0.026
Residual	9	2443.449	271.494		
Total	10	4352.909			

$SSR = 1909.460$, $SSE = 2443.449$, and $SST = 4532.909$. As expected, $SST = SSR + SSE$. Hence, the percentage of the total variation in the observed y-values that is explained by the regression is

$$\frac{SSR}{SST} = \frac{1909.460}{4352.909} = 0.439.$$

PROBLEM 14.5 Excel output:

=CORREL(HEIGHT,WEIGHT) yields 0.662

Lesson 15

PROBLEM 15.1
a) From the Excel output at the top of this page, s = 16.48; thus we have, $s_e = 16.477$.
b) 16.477 lb.

PROBLEM 15.2

Excel Residual Plot:

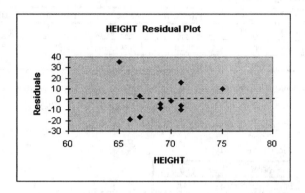

Excel Probability Plot:

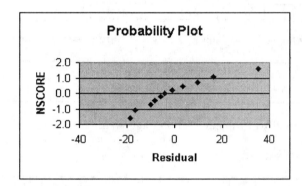

PROBLEM 15.3

Excel partial regression output:

	Coefficients	Standard Error	t Stat	P-value	Lower 95%	Upper 95%
Intercept	-174.493	126.260	-1.382	0.200	-460.114	111.128
HEIGHT	4.836	1.824	2.652	0.026	0.711	8.962

From the figure, we see that the P-value is 0.026. The data provide sufficient evidence to conclude that the slope of the population regression line is not zero; hence height is a useful predictor variable for weight.

PROBLEM 15.4

From the display in the preceding problem, we can be 95% confident that β_1 is between 0.711 and 8.962 pounds.

PROBLEM 15.5

From the output on the following page, we can be 95% confident that the mean weight for the population of males 18–24 years old who are 70 inches tall is somewhere between 152.31 and 175.79 pounds.

Excel **Estimating Means and Predicting Values** panel:

α =	0.05	Estimating Means and Predicting Values			
Confidence	95.0%	Confidence	Interval	Prediction	Interval
HEIGHT	Fit	Lower Limit	Upper Limit	Lower Limit	Upper Limit
70.00	164.048	152.313	175.782	124.971	203.125

PROBLEM 15.6 From the above output we can be 95% certain that the weight of a randomly selected male 18–24 years old who is 70 inches tall will be somewhere between 124.97 and 203.13 pounds.

PROBLEM 15.7 Excel **Correlation t-Test** output:

	Correlation	t-Test	
DATA FILE	α =	0.05	
NAMES:	Alt Code	1	P-value
HEIGHT	Correlation	t-statistic	0.013
WEIGHT	0.662	2.652	Reject

Evidently, the variables, height and weight are positively correlated.

Lesson 16

PROBLEM 16.1 See the solution to Problem 11.5 on page A-13.

PROBLEM 16.2 Excel **ANOVA** output:

ANOVA						
Source of Variation	SS	df	MS	F	P-value	F crit
Between Groups	115.111	2	57.556	8.534	0.003	3.682
Within Groups	101.167	15	6.744			
Total	216.278	17				

With a P-value of 0.003, reject H_0. The data provide sufficient evidence to conclude that mean sales resulting from the different policies are not all the same. From the same output, $SST = 216.278$, $SSTR = 115.111$, and $SSE = 101.167$.

Confidence intervals for each of the three policies may be read from the following output obtained from the confidence interval panel in the **ANOVA** worksheet.

df	MS	α	
15	6.744	0.05	
95.0%	Confidence	Intervals	
Groups	Lower Limit	Upper Limit	
POLICY1	20.240	24.760	
POLICY2	20.907	25.426	
POLICY3	25.907	30.426	

INDEX

Add-ins, 13
Alphanumeric data, 8
Alt key, 4
Alternative hypothesis, 97
   choice of, 97
Analysis of variance, 170
   in regression, 155
ANOVA
   data analysis tool, 172
   assumptions, 171
   confidence intervals, 175
   identity, 173
   single factor, 172
   sums of squares, 174
Association, 141
AutoSum, 11, 33
AVERAGE function, 31

Bar,
   formula, 4
   menu, 3
   scroll, 4
   status, 4
   task, 4
   title, 3
   tool, 4
Bernoulli trials, 59
Between groups sum of squares, 172
Bin, 21
Binomial distribution, 59
   plot of, 61
   Poisson approximation, 65
Binomial probability formula, 59
Binomial random variable, 59
   mean of, 61
   standard deviation of, 61
BINOMDIST function, 59
Box,
   check, 6
   dialog, 6
   name, 4, 9
   text, 4, 6
Boxplot, 40
Button,
   cancel, 7
   close, 3
   collapse dialog, 9
   control, 3
   copy, 49
   down triangle, 6
   maximize, 3
   OK, 7
   option, 7
   paste, 49
   radio, 7
   restore, 3
   scroll, 4

Cells, 4
   range, 8
Central limit theorem, 84
Central tendency, 31
Chart
   output, 24
   wizard 61, 64, 151
Check box, 6
CHIDIST function, 118, 138
CHIINV function, 117, 137
Chi-square curve, 117, 137
   χ^2_α, 117, 137
Chi-square distribution, 117, 137
Chi-square goodness-of-fit test, 138
   test statistic for, 140
Chi-square independence test, 145
   test statistic for, 147
Chi-square procedures, 136
Classes, 21
Clicking a mouse, 5
Coefficient of determination, 153
Collapse dialog button, 9
Column, 8
Completely randomized design, 172
Conditional
   distribution, 50
   probability, 50
Confidence interval for a population mean
   in one-way analysis of variance, 174
   in regression, 164–169
   normal population and σ known, 90
   normal population and σ unknown, 92
Confidence interval for a population standard deviation, 121
Confidence interval for one population proportion, 129
Confidence interval for the difference between two population means
   normal differences and paired samples, 112
   normal populations with equal standard deviations, using independent samples, 107
   normal populations with unequal standard deviations, using independent samples, 109
Confidence interval for the difference between two population proportions, 132
Confidence interval for the ratio of two population standard deviations, 123

Confidence interval for the slope of a population regression line, 164
Contingency table, 45, 141
Copy button, 37, 49
Corrections, 11
Correlation, 156, 167
Correlation coefficient
  linear, *see* Linear correlation coefficient
CORREL function, 156
COUNT function, 34
COUNTIF function, 57
Cumulative probability, 59
Cutpoints, 21

Data Analysis Toolpack, 13, 20
Degrees of freedom
  for a χ^2-curve, 117, 137
  for an F-curve, 123, 171
  for a t-curve, 92
Descriptive statistics tool, 38
Descriptive measures
  of central tendency, 31
  of spread (variation), 35
Deselect, 6
Dialog box, 6
Discrete random variable, 54
  mean of, 58
  probability distribution of, 56
  standard deviation of, 58
Dotplots, 27

Ellipses dots, 6
Enter key, 8
Equally likely, 44
Error sum of squares
  in one-way analysis of variance, 173
  in regression, 153
Escape key, 7
Excel 97, 2
  corrections, 11
  file, 12
  input/output, 8
  leaving, 13
  logo, 2
  menus, 5
Mode, 32
MODE function, 33
Mouse, 5
  pointer, 4
Expected frequencies, 140
Explained variation, 154
Exponential curve, 85
Exponential distribution, 85
Extension (filetype), file, 12

F-distribution, 123, 171
  F_α, 123, 171
FDIST function, 124
FINV function, 124
Files, 12
  extension (filetype), 12

workbook (.XLS), 12
Five-number summary, 38
f/N rule, 44
Frequency
  distribution, 24
  histogram, 24
Functions
  AVERAGE, 59
  BINOMDIST, 59
  CHIDIST, 138
  CHIINV, 137
  CORREL, 156
  COUNT, 34
  COUNTIF, 57
  FDIST, 124
  FINV, 124
  MAX, 35
  MEDIAN, 32
  MIN, 35
  MODE, 33
  NORMDIST, 73
  NORMINV, 74
  NORMSDIST, 70
  NORMSINV, 71
  POISSON, 63
  ROUND, 17
  STDEV, 36
  SUM, 33
  SUMPRODUCT, 59

GOF panel, 139
Graphs (charts), 24
Grouped-data table, 20
Grouping data, 20
  using classes based on a single value, 22

Help, 5
Histogram,
  frequency, 21
  for single-value grouping, 22
Histogram tool, 20
Hypothesis test, 96
  choosing the hypotheses, 97
Hypothesis test for a population linear correlation coefficient, 168
Hypothesis tests for a population standard deviation, 121
Hypothesis test for one population mean
  σ known, 99
  σ unknown, 102
Hypothesis test for one population proportion, 131
Hypothesis test for statistical dependence of two characteristics of a population, 145
Hypothesis test for the slope of a population regression line, 162
Hypothesis test for two population means
  large, independent samples, 138
  normal differences and paired samples, 113
  normal populations with equal standard deviations, using independent samples 107
  normal populations with unequal standard deviations, using independent samples, 109

Hypothesis test for two population proportions, 132
Hypothesis test for two standard deviations, 124
Hypothesis testing
  P-value approach to, 98
  confidence interval approach to, 98

Icons, 4
Independent samples, 107, 109, 132, 166
Intercept, 153
Interquartile range, 37

Joint probability distribution, 48

Least-squares criterion, 150
Leaving Excel, 13
Left-tailed test, 97
Lessons, 2
Linear correlation coefficient, 156, 167
Linearly uncorrelated, 156

Marginal distribution, 48
Marquee, 9
MAX function, 35
Mean
  of a binomial random variable, 61
  of a discrete random variable, 58
  of a Poisson radom variable, 62
  of a population, see Population mean
  of a sample, see Sample mean
  sampling distribution of, 59
  standard error of, 39
Measures of center (central tendency), 31
Measures of spread (variation), 35
Median, 32
MEDIAN function, 32
Menu selection, 5–8
Menu bar, 3
Midpoint, 21, 23
MIN function, 35

Naming cells, 8, 9
Negatively linearly correlated, 156
Normal curve
  finding areas under, 70, 73
  finding the value(s) for a specified area, 71, 74
  parameters, 68
  standard, 70
Normal differences, 113
Normal probability plots, 74
Normally distributed random variables, 68
  simulating, 68
Nscores, 75
Null hypothesis, 97

Observed frequencies, 139
One-way analysis of variance, 171
  assumptions for, 171
  test statistic for, 173
One-way ANOVA identity, 173

One-way ANOVA table, 173
Outliers, 77

P-value, 98
Paired
  differences, 113
  samples, 112
Parameters,
  binomial, 59
  exponential, 85
  normal, 68
  Poisson, 62
Paste button, 37, 49
Patterned data entry, 23
Pearson product moment correlation coefficient, see
    Linear correlation coefficient
Pivot Table Report, 50, 146
Poisson distribution, 62
  approximating the Binomial distribution, 65
  mean of, 62
  plot of, 64
  standard deviation of, 64
POISSON function, 62
Pooled standard deviation, 107, 169
  in analysis of variance, 169
Population, 41
Population linear correlation coefficient, 167
Population mean, 41
Population proportion, 129, 132
Population regression line, 162
Population standard deviation, 119, 126
Positively linearly correlated, 156
Prediction interval, 166
Predictor variable, 151
Printout, 16, 17
Probability, 44
Probability distribution
  binomial, 59
  chi-square, 117, 137
  conditional, 50
  discrete, 54
  F, 123, 166
  interpretation of, 56
  joint, 48
  marginal, 49
  normal, 67
  Poisson, 62
  standard normal, 70
  t, 92
Probability plot,
  see NSCORES
Proportion,
  population, see Population proportion
  sample, see Sample proportion

Quartiles, 37

Random Number Generation tool,
  simulating a discrete distribution with, 52
  simulating an exponential distribution with, 86
  simulating a normal distribution with, 68

Random sample, 14
  simple, 14
  systematic, 16
Random sampling option, 15
  systematic, 16
Random variable, 54
  binomial, 59
  discrete, see Discrete random variable
  normally distributed, 73
  Poisson, 62
Range, 8, 35
Regression tool, 152
Regression equation, 150
Regression inferences
  assumptions for, 158
Regression sum of squares, 155
Relative frequency, 20
Residuals option, 160
Residuals in regression, 159
Response variable, 151
Right-tailed test, 97
ROUND function, 17

Sample mean, 34
Sample proportion, 130
  pooled, 133
Sample standard deviation, 35
Sampling
  random, 14
  tool, 14, 15
  with replacement, 14
  without replacement, 14
Sampling distribution of the mean
  for general populations, 85
  for normally distributed populations, 82
Sampling error, 80
SAVE icon, 14
Scatter diagram (plot), 151
Select button, down triangle, 6
Selecting menus, 5–7
Simple random sample, 14
Simulation
  discrete distribution, 52
  exponential distribution, 86
  normal distribution, 68
Single-value grouping, 22
  histograms for, 26
Slope of a regression line, 162

SORT button, 38
Standard deviation
  of a binomial random variable, 61
  of a discrete random variable, 58
  of a Poisson random variable, 64
  of a population, see Population standard deviation
  of a sample, see Sample standard deviation
Standard error
  of the mean, 39
Standard error in regression, 158
Standard normal curve, 70
  areas under, 70
  finding the z-value(s) for a specified area, 71
Statistical software packages, 2
STDEV function, 36
Submenus, 5, 6
Successes, 59
SUM function, 33
SUMPRODUCT function, 59
Systematic sample, 16

Tab key, 7
Table, contingency, 62, 141
  split bar, 14
t-distribution, 92
  t_α, 92
Testing normality, 74
Text box, 4, 6
Title bar, 3
Total sum of squares,
  in one-way analysis of variance, 173
  in regression, 155
Two population proportions, 132
Two-tailed test, 97

Uncorrelated, 156

Variation (spread), 35

Within groups sum of squares, 172
Worksheet, 4
Worksheet file, 16

y-intercept, 153

z_α, 72